« THE BIRTH OF COMPUTER VISION »

The Birth of Computer Vision

JAMES E. DOBSON

University of Minnesota Press
Minneapolis
London

The University of Minnesota Press gratefully acknowledges
the financial assistance provided for the publication of this book
by Dartmouth College.

Published by the University of Minnesota Press
111 Third Avenue South, Suite 290
Minneapolis, MN 55401-2520
http://www.upress.umn.edu

ISBN 978-1-5179-1420-2 (hc)
ISBN 978-1-5179-1421-9 (pb)

Library of Congress record available at
https://lccn.loc.gov/2022040925

Contents

Acknowledgments vii

Introduction 1

1 Computer Vision 27

2 Inventing Machine Learning with the Perceptron 61

3 Describing Pictures: From Image Understanding to Computer Vision 97

4 Shaky Beginnings: Military Automatons and Line Detection 133

Coda 165

Notes 177

Index 201

Acknowledgments

The writing of this book was completed during the first few months of the Covid-19 pandemic. I owe thanks to many friends and colleagues with whom I was able to find comfort and community during the most lonely days of the months and years that followed. Thank you especially to Timothy Baker, Sarah Coulter, Jason Houle, Johanna Mirenda, Alan Taylor, Melanie Taylor, and Louis A. Renza. Thank you also to my many friends in Maine, including Dwain Clough, James Gammon, Cody LaMontagne, Doug and Kris Stahle, and Dick Perkins.

Two of my colleagues in the English and creative writing department at Dartmouth College provided essential commentary and advice. In the midst of all the chaos and uncertainty during the pandemic, Donald E. Pease provided a careful reading of my argument and material through its several different shapes and forms. He also gave me an opportunity to present earlier versions of some of this material as a plenary lecture during the Futures of American Studies Institute. Aden Evens has an unmatched ability to think through both the technical and theoretical concerns raised by this project and was instrumental in helping me reshape and reconfigure key moments in this book. My work and thinking about all matters digital and critical is much stronger thanks to the ongoing generosity of spirit and critical questions posed by Aden. Thank you to my other colleagues in English and creative writing, especially Colleen Boggs, Peter Orner, Patricia Stuelke, Vievee Francis, George Edmondson, Alexander Chee, Melissa Zeiger, Michael Chaney, Matthew Olzmann, and Andrew McCann, as well as to Matthew Ritger and Jessica Beckman, who arrived in the early days of the pandemic and have become great friends.

Dartmouth College and the dean of the faculty have been unwavering in their support of my work and career over the past several years. Thank you to Elizabeth Smith, Barbara Will, Dennis Washburn, Matthew Delmont, and Samuel Levey.

The Futures of American Studies Institute continues to be one of the most important venues for my work. Thank you to institute codirectors and regulars including Duncan Faherty, Eric Lott, Elizabeth Maddock Dillon, Donatella Izzo, and Soyica Diggs Colbert.

I owe an enormous debt of gratitude to Mark D. Koch, who has helped in numerous ways to make sure that the Institute for Writing and Rhetoric has become successful through the challenging pandemic terms and who has been a dear friend and colleague.

Thank you to Doug Armato and Zenyse Miller at the University of Minnesota Press. Thank you, as well, to Karen Hellekson for the careful copyediting. I would also like to thank David Gunkel for his incredibly helpful feedback, recommendations, and commentary on this project.

Finally, I owe an enormous debt to Rena J. Mosteirin, my frequent collaborator and wife. Thank you, Rena, for your support of this and every other project, and most especially for your love.

Introduction

> It is seeing which establishes our place in the surrounding world; we explain that world with words, but words can never undo the fact that we are surrounded by it. The relation between what we see and what we know is never settled.
>
> —John Berger, *Ways of Seeing*

In the twenty-first century, we have witnessed the emergence of new forms of computerized surveillance systems, high-tech policing, and automated decision-making systems that have increasingly become entangled, functioning together as a new technological apparatus of social control. The movement of data from this apparatus into governing and juridical practices has become a pressing concern as we learn more about how the widespread use of these systems has disproportionately targeted and disenfranchised individuals and classes of people, in particular Black and poor people. Virginia Eubanks argues that the high-tech, data-driven tools involved in these new systems function primarily to automate already existing social inequalities under the cover of what she calls the ethical distance between those who are subjected to these systems and the professional middle-class public, who have thus far avoided the algorithmic gaze.[1] Eubanks provides a narrow account of automated decision-making that considers mostly numerical data applied in situations in which individualized services are only available at a premium cost; she thus sees these decision-making procedures as primarily directed toward the poor. Computer-enabled automation, Eubanks argues, preserves human interaction for those with access to privilege and produces distance between these people and the subjects of automated decision-making systems. This distance-creating capacity of computerized tracking and decision-making has been key to its widespread adoption in everyday governing systems. In computerized systems, especially in applications situated in a machine learning context, distance leverages the gap between sample and target, training and application, to render abstract what on closer inspection would otherwise be seen as concrete and contestable.[2] Distancing gives the appearance of objectivity by reinstating what has long been a major trope of modernity, the disinterested observer, through the fiction of an autonomous computer vision. The ideological move by which governments, police, and corporations present their high-tech tools as distant, objective viewers of data requires historical critique to understand the ways in which these tools have functioned and continue to function as what Safiya Umoja Noble terms "algorithms of oppression."[3] Most systems that comprise the contemporary surveillance state through the widespread deployment of what Shoshana Zuboff usefully terms "surveillance capitalism" are primarily visual.[4] These systems operate on images acquired from the growing streams of surveillance video, from policing and judicial photographic records, and increasingly from personal image data shared by everyday citizens on the web through the increasing number of image- and video-based social media applications as well as video acquired through home security and monitoring devices, stored and disseminated via personal computers and phones. The widespread use of these technologies and their connection to surveillance have changed how we understand images as well as our movement through space in both public and private contexts. These advanced technologies of seeing are thus much closer observers than we might initially think.

Computer vision algorithms are frequently invisible; they may seem esoteric or too highly specialized to give them much serious thought, but these technologies are now utterly pervasive. They run in the background, as it were, seeing and increasingly determining the world. These technologies are not just imposing order on the content of images; as prosthetic visual technologies, they are also changing how we see. In the twenty-first century, the distinction between human and computer vision has become increasingly hard to define. This is no accident. These technologies were initially developed in part to help researchers understand human perception. In modeling perception, such technologies provided a new set of metaphors that enabled a redescription of physiological systems. They are more than just models of human vision; they are designed to exceed our visual capacities. Computer vision can make the invisible visible. It can detect the presence of obscured objects. It can augment reality, identifying danger and tracking movement. It can learn from the past to add structure to the present. It can also replicate biases. Most crucially, the birth of computer vision inaugurated new techniques of governance and regimes of visuality that reconfigure our conceptions of perception and observation. These algorithms have become domesticated, in multiple senses of the word. While many of the core computer vision algorithms were initially developed for tracking and predicting the actions of foreign armies, they are now used in digital photo albums, are widely found in social media applications, and are even embedded in automobiles.[5]

It makes no sense, however, to discuss computer or machine vision in the abstract as ahistorical operations because these algorithms and procedures, like almost all computational methods, exist in temporal flux, sometimes undergoing reorienting revolutions and other times more subtle transformations. Complex algorithms rarely sit still. This book traces the historical development of these new modes of computerized perception and argues that the social and technological conditions of the mid-twentieth century that shaped their invention and their subsequent transformation from their birth continue to influence the pervasive technologies of seeing in the twenty-first century. The readings of algorithms that follow are drawn from methods inspired by the discourse of computer vision itself. I use these methods to interpret several major computer vision algorithms by iteratively following the steps of the solutions generated by early attempts to automate perceptual tasks. In several of the primary (if not primal) scenes and imaginative scenarios that motivated the birth of computer vision, we see a number of machines and discrete computational units altering data and making decisions, but we can also identify a cast of characters, including the operators of these imagined seeing machines, the knowledge workers whose expertise is proposed to be extracted and automated, and the observers of these scenes themselves—that is, the inventors and parents of computer vision.

One major goal of this book is to deconstruct the absolute

division that has been drawn between accounts of human and machine vision. The invention of computer vision has produced a displacement of the observing human subject, but this displacement did not remove the human perceptual subject from computer vision. Early accounts of computer vision especially deploy fantasies of a completely autonomous mode of perception and at the same time depend on extracting knowledge from prior perceptions, even if these are composite and no longer indexed to a single observer. Any prospects for a machine vision, as I will argue in the following chapters, exist only as a result of the encoding of preexisting knowledge and perceptions that are imagined, framed, experienced, and described by humans. In short, human perception is the precondition for machine vision. Despite the reference to automaticity that saturates computer vision discourse, in both the past and present, actual perceiving human beings teach and train the larger systems in which they are embedded. This provides what we might call first sight, a mode of seeing that precedes and persists in computer vision. Encoded knowledge, prior decisions, and naming schemes derived from these highly subjective and historically contingent human activities become precepts, the criteria and rules by which a machine can operate in the absence of an operator. Within computer vision algorithms, which are almost always attached to learning algorithms, knowledge in the form of rules learned from the past is used to make decisions in the present. The selection criteria through which norms were established in the past haunt the production of knowledge in all learning systems. It is this sense of the historicity of computer vision, the multiple ways in which the past is preserved as generalizable rules for recognition in the present, that motivates the following genealogical analysis of computer vision as a field and practice.

These changes in perception brought into being by computer vision are akin in scale and cultural importance to an earlier moment in the history of vision: the epistemic shifts produced by the use of visual technologies in the eighteenth and nineteenth centuries. As Jonathan Crary argues, these earlier seeing technologies, which include the camera obscura, the stereoscope, and film photography, had dramatically different epistemic effects. The former two devices made vision more subjective, whereas the latter produced a rupture in the scene of observation: "Photography had already abolished the inseparability of observer and camera obscura, bound together by a single point of view, and made the new camera an apparatus fundamentally independent of the spectator yet which masqueraded as a transparent and incorporeal intermediary between observer and world."[6] Computer vision inserts itself in the scene of observation, taking the place previously occupied by the camera itself. It could thus be understood as an acceleration of the project to displace the human perceptual subject and the seeing eye.[7] What we see when we see the output of computer vision, if this output is even rendered in image form, is an impossible perspective for human vision. It is impossible, as this book will argue, because these data are produced from, among many other possible reconfigurations of image data, overlays or other combinations of images and segments of images, a complex or warped coordinate system, averaging and other statistical manipulations of images. Computer vision also makes use of what we might think of as a mobile frame, a variable positioning of the camera and computation—one such example of a highly mobile frame is the aerial view of a small and nimble aerial drone—that cannot be taken up by humans. The image data framed by computer vision belongs to another space, a space no longer representing a possible point of view for humans. Nonetheless, computer vision asserts itself as superior, and users of these technologies are encouraged to identify with this impossible gaze.[8] Take, for instance, backing up a car. When a driver puts a car in reverse, the look is no longer directed backward but rather forward, to a display mounted on the dashboard, with lines and measured distances overlaying a real-time photographic image. The technologies active in this and similar scenes, which are everyday encounters with computer vision, build on prior technological developments in distant seeing and observation, primarily through aerial photography and its altered perspective—a perspective that is at the center of much computer vision research and that uses its secure position to turn the entire visual field into a mappable and ultimately knowable space.

The perceptual and ontological shifts brought into being by computer vision are important and far reaching, especially those appearing during this technology's initial articulation and development in the mid-twentieth century. This early period in the history of computer vision created many of the algorithms that continue to be used in the present. A field of research dedicated to the computational understanding of images and vision was also simultaneously solidified. However, more importantly during this period, from the late 1950s until the mid-1970s, computer vision underwent two crucial ontological shifts that produced changes in the understanding of vision and visual culture. These shifts altered the possible meaning of images from the perspective of the algorithms and introduced changes in the imaginary scene of computer vision, including what might be called the situatedness of the machine and its aperture in relation to the visual scene. The first version of computer vision involved the comparison of entire images and pixel intensity values. The image concept within computer vision at the time was understood to be in an isomorphic relation with the photographic image—that is, it was comparable, even if reduced in resolution, or identical in terms of the information content. The first ontological shift changed the source of knowledge about images from the entire image to a reduced set of information produced by feature-orientated algorithms. This shift involved a fracturing of the image into different regions that corresponded to distinct and preidentified image features. The second shift involved a reconfiguration of the scene of scene analysis, from the scanning and sensing of previously acquired image data to what might be called sightless seeing that joined the visual field with its site of analysis. This book describes an ontological shift from lower-level, pixel-based picture data acquired directly from scanners or sensors to higher-level symbolic representations produced through feature detection systems within early computer vision. This reconfiguration, however, was not absolute, and prior methods were not completely obsoleted. Computer vision techniques used in this later phase, after the second ontological shift, would continue to make use of pixel-level transformations, perhaps before the extraction of features as a form of preprocessing to normalize or reduce complexity, or maybe after computation on these features in order to place them back in the original image space. Many contemporary basic image-processing techniques continue to operate on the entire image. With the present revitalization of machine learning techniques, including deep neural networks, some of the most popular algorithms operate on all available pixels without knowledge of higher-level image features.

The developments in computer vision analyzed in this book are intertwined with those in psychology and physiology. The shifts identified are also caused by historically specific technical limitations as well as insurmountable philosophical problems. At stake is an understanding of the difference between seeing and perception, the core problem of the second half of this book and the period under discussion. It would not be incorrect to say that the deep wish of computer vision is to render vision fully objective and detached from human perception. Machine perception is thus the next step in the continuation of the regimes of visuality that were launched by sites and technologies, from the invention of the (mostly imaginary) bird's-eye perspective to the panopticon. Yet while information extraction tasks like categorization or identification are the priority, computer vision also produces new images for human consumption. These images, especially during these techniques' period of development, are frequently presented as enhanced images with supplemental information such as the ubiquitous boxes drawn with lines surrounding objects of interest and paired with text labels identifying visual objects.

The various visual technologies developed over the years, especially those finding use in or as scientific instruments, have long been critiqued for making unwarranted claims to objectivity. Donna Haraway addresses a set of scientific visual technologies as a form of situated knowledge in her feminist intervention in science studies. For Haraway, being situated means acknowledging the limitations of all knowledge and the mediation involved in any claim to objectivity. One source of knowledge acquisition of particular importance to Haraway's argument is instrumental vision. "I would like to insist," Haraway writes, "on the embodied nature of all vision and so reclaim the sensory system that has been used to signify a leap out of the marked body and into a conquering gaze from nowhere."[9] In critiquing the notion of an unmediated and objective scientific gaze, Haraway wants to destabilize the claim of "seeing everything from nowhere" that she sees as active in scientific accounts.[10] She identifies a wide range of visual instruments as participating in logics of disembodiment, including "sonography systems, magnetic reasonance *[sic]* imaging, artificial intelligence–linked graphic manipulation systems, scanning electron microscopes, computed tomography scanners, color-enhancement techniques, satellite surveillance systems, home and office video display terminals, cameras for every purpose."[11] In its early years, and to some degree still today, computer vision was propped up by the desire for an objective perspective on the world.

The imaginary gaze of computer vision is constructed by obfuscation and abstraction; we can say that it achieves whatever claims to objectivity and authority it has by denying its location, its historicity, and its situatedness. By functioning as tools of domination and control, and linked as they are to imperialist desires to name and categorize, such seeing technologies map space and territories. For feminist scholars, invoking the body is a critical maneuver when that body is still present but obscured behind an instrument. Within the normative paradigm of computer vision, that body has seemingly been entirely displaced from the scene of observation. Yet computer vision is just as historically determined and situated as the other seeing technologies Haraway invokes—even more so, perhaps, because of the deep embedding of prior subjective judgments involved in these very displacements of the observing human subject. The algorithms and methods of computer vision matter because congealed within these abstract operations, we discover encoded thinking about the relationship between subjects and objects and what it means to see and perceive objects in the world. Understanding the development of these technologies is a concern of not just computer history but also intellectual history.

Critical Algorithm Studies

This book is primarily a contribution to the field of critical algorithm studies. It is an extended cultural studies reading of a specific subset of algorithms that were essential to the development of computer vision research projects and to the wide variety of today's most commonly used image and object recognition applications. In order to analyze these algorithms, however, a richer vocabulary and framing are needed than those found in typical accounts of computer technology. Because these algorithms are used as substitutes for human vision through the modeling of digital or digitized images that are understood as representations of the real world, some theorization of the image, observation, and perception is necessary, as specific understandings of these concepts are embedded in the algorithms and in their use. These algorithms must be understood as possessing a historicity of their own such that unpacking them requires reading them in relation to the discursive moments that made them possible. At the same time, these algorithms also need to be connected to longer histories of visual culture, scientific objectivity, and the surveillance techniques that precede them, determine their reception, and shape their ongoing use. Understanding the histories and theories of vision and machine perception that are congealed, inherited, reproduced, or even contested by contemporary algorithms is a crucial component of a reflective use of computer vision.[12]

The computer vision algorithms that I examine in this book provide much of the predictive power behind contemporary face and object recognition applications. These applications have become increasingly common. Versions of these same algorithms come preinstalled on phones and computers. They are used by all levels of government as well as by nongovernmental organizations, including the corporations that manufacture our devices and software. Most contemporary critical work on computer algorithms has focused on what we might call the phenomenological experience of digital devices—that is, what we experience of the digital world when we use computers, smartphones, and other technological artifacts. This approach enables critics to provide readers with a sense of the different ways in which they might interact with computers and other technological devices and how these devices address human users. The phenomenological method locates the critical site of inquiry in what is called human–computer interaction and probes beneath the surface to understand what we see, hear, and touch as we make use of computing devices. Take, for example, a critique of the Google search application and its associated black box algorithms through the results returned from a query, or the results returned from a voice-activated interaction with a smart appliance and information retrieval system such as Amazon's Alexa or Apple's Siri.[13] The dominant paradigm used for understanding technologies, especially historical technologies, remains the theory of what is called the "social construction of technological systems" emerging from science and technology studies.[14] This approach understands technology not necessarily just in material form, as particular devices or artifacts, but also as embedded in much larger systems. Other critical accounts of algorithms, influenced by science and technology studies and cultural studies, locate these objects within the larger social forces and structures in which they were created and operate. These approaches have focused attention on the ways in which algorithms enable or enact whole ranges of suspect ideological assumptions when deployed as a component of a social system that might include numerous biases, articulating explicitly or implicitly sexist, racist, and other undesirable historical or dominant norms.[15]

Computer vision names both a subfield in the academic study of computer science and a rapidly changing area of applied software development that reaches into many aspects of contemporary everyday life. Despite their origins in the laboratories of psychologists, the secretive spaces of Cold War–era university research facilities funded by the U.S. Department of Defense, and the offices of early Silicon Valley corporations, these technologies are an inseparable part of twenty-first-century visual culture. The cultural work of computer vision takes place at the intersection of theory and practice. For the most part, these algorithms share an understanding of the ontological status of images; they also share a deep and complicated history. Many of these algorithms were initially developed by a small group of engineers and computer scientists with a shared sense of purpose and goals shaped by funding agencies. The algorithms that make up computer vision applications are now some of the most widely used in contemporary digital life. These applications are central to what we might think of as the tool kit of distancing techniques in computerized governing systems. They detect our faces when we glance into a camera as we cross national borders; they keep us in our lane by locating the strips of paint on the highway; they recognize our license plates as we drive without stopping at tollbooths. Computer vision algorithms assist in focusing smartphone lenses to better capture and automatically extract faces, then matches these faces with those already extracted and known faces in our digital photo albums and social media applications. These same algorithms detect the presence of barcodes (or, more commonly now, QR or Quick Response codes) on signs and products in grocery stores and at restaurants. Computer vision applications enable us to unlock phones and laptops without a password when we point the devices at our faces. The successful everyday use of these sorts of computer vision applications has made people accustomed to and even accepting of these technologies, although there is growing recognition that these algorithms might be too powerful and too widely used.[16] By using a series of important computer vision algorithms as case studies, this book examines the creation and ongoing development of these concepts and practices. Drawing on the resources of critical theory and cultural studies, this book presents an interdisciplinary approach to understanding the historical development of computer vision through the initial goals, stunning achievements, and worrisome aspects that linger and are reproduced within the field and its products.[17]

Computer vision methods are often paired with machine learning and artificial intelligence algorithms because they were developed and articulated alongside each other. What we now recognize as computer vision applications informed the design of many popular and important machine learning algorithms. They provided the benchmarks, milestones, and impressive demonstrations by which academic researchers, military contractors, and commercial developers proved the capabilities of their inventions and patented techniques. Image analysis and recognition, basic and simple tasks for human beings, represented the boundaries of computation in the 1960s and 1970s. It is hard to imagine now, but computer vision preceded the ability to see a display or printout of an image; image analysis took place within the dark room of computation. Today, similar applications produce awe and wonder with their higher resolution and better accuracy, yet these computer vision applications remain tied up with complicated code and confusing, complex models. Interpreters of these models have lost much of their prior ability to understand the criteria by which decisions are made about the classification of visual objects. Even more troubling is the use of these systems in everyday life, especially in policing, where they have become a source of much anxiety in an image-saturated culture that has continued to criminalize the mere presence in public of certain bodies.

Genealogies of Seeing Technologies

The dubious and frequently harmful projects of computer vision, especially those addressing human bodies and faces, have a long history. Despite the ideological belief in ceaseless forward progress in technology, older methods and algorithms are not obsoleted in order to be replaced by newer and better methods; instead, they are often retained in slightly different forms.[18] The field of computer vision and its applications might be said to be haunted by prior visions and visionaries. It is not just the explicit design goals and project imperatives that served as conditions of possibility for developed technologies but also the prejudices, assumptions, and biases of the programmers and project managers that reside in the produced code and related discourse. These ideologies shape the affordances of technical objects and the ways in which humans interact with technology. Norms, beliefs, and assumptions are frequently baked into the basic operations of technology; sometimes, in using these systems, users invisibly iterate and operationalize these norms, while at other times these ideologies are suspended in the background, as it were, waiting to shape unknown and unpredicted future uses. One example is the command and control systems created and supported by U.S. Department of Defense funding agencies. These systems needed automated object recognition systems to analyze the rapidly growing number of military surveillance photographs acquired during the Vietnam War. They also required new forms of object detection, from ship and tank identification to face detection. From the late 1950s until the early 1970s, scientists and engineers, in exchange for the funding of their research programs and students, brought the desires and motives of the newly created military-industrial complex into their universities and labs, shaping the field's foundations as well as many of today's most important algorithms.[19] The resulting technologies have become central to the predictive operations of what Brian Massumi calls ontopower, the operationalization of the logics of preemption that takes as its target human perception itself.[20] Tracing the development of computer vision from these labs to the present allows us to understand how our changing ways of seeing are struck through with the residues of the past.

The statistical techniques used to enable machine learning were invented when these algorithms were materially embedded in special-purpose computing machines, which were frequently analog rather than digital devices. When we see popular cultural representations of artificial intelligence in a general robot or android form, something important about the origin story of contemporary computing is manifested in these autonomous mechanical caricatures. Machine learning appeared when general-purpose computers were still new on the scene and not yet able to interface with custom hardware. The available programming languages were not yet adaptable to some of the requirements identified as crucial to advanced statistics and the analysis of pictures. Alternative computing devices that simulated existing accounts of visual perception were developed to sense photographic images, convert them into computable signals, and learn to recognize patterns from these signals. Because these new machines were understood to be self-organizing through the contemporary discourse of cybernetics, they were imagined as much more autonomous than they really were. The language of the machine and the automaton was incorporated into the discourse surrounding these new learning technologies, and when general-purpose computers could handle these tasks, a few short years after the initial invention of some of these methods, the phrase "machine learning" stuck. To tell the story of the development of computer vision, the ongoing debates and disagreements in the discourse of computer vision and machine learning, the algorithms themselves, the materiality of these algorithms and their implementations in specialized hardware, and the history of the labs, funding agencies, and researchers are all required.

These algorithms, devices, and methods were initially developed during a twenty-year period beginning in the last years of the 1950s and ending in the late 1970s. During this moment, computer vision was first defined and many of the most popular contemporary methods were invented. It is a crucial period because the ontology of the image within computer vision underwent important transformations, with shifts in the understanding of perception shared with other adjacent research fields, including neurophysiology. This period is also linked to the explosive growth in the use of computers by businesses, universities, and the military. Computing during this time was divided into research and academic computing, and business computing. This period roughly corresponds to what computer historian Paul E. Ceruzzi calls the go-go years, a moment he characterizes as dominated by the development of commercial computing and the explosive growth of IBM and their System/360 computer. As was typical of the few computer companies at the time, IBM focused on providing an entire computing environment that was characterized by "large, centralized mainframe installations, running batches of programs submitted as decks of punched cards."[21] Ceruzzi's analysis of the System/360 environment demonstrates the degree to which these mostly business-oriented data processing systems obtained a monopoly on computing once they were installed. Research environments, however, were characterized by a greater diversity of computing systems. The labs and centers that were the origin sites of computer vision algorithms—SRI International in Menlo Park, California; Cornell Aeronautical Laboratory in Buffalo, New York; and Brookhaven National Laboratory in Upton, New York—made use of commercial computing systems from vendors like IBM, Digital Equipment Corporation (DEC), and Honeywell, but also more specialized computers, including Scientific Data Systems' SDS 940. Custom hardware was also invented and produced at these labs—hardware that supplemented existing commercially available digital computers, and in some cases provided an alternative model of computation.[22] The researchers developing computer vision methods made use of a variety of computer systems and programming languages. This heterogenous computing environment separates these research sites from the majority of computer installations during the 1960s and 1970s in which a single vendor supplied most computing power in a manner that Ceruzzi, riffing on IBM's own marketing, terms "full circle" computing.[23]

The approach that I take in this book might be considered a specialized and specific form of the history of ideas or intellectual history. Algorithms, as this book will argue, are primarily ideas. Created and programmed in software and/or hardware, they are artifacts shaped by the material and historical contingencies of their moment of genesis. But at their core, they are abstract conceptualizations of problem-solving tasks. The developed solutions to these problems draw on the resources of prior solutions and evolve in response to changing historical conditions. Algorithms frequently have specific or reference implementations; they are developed alongside historical limitations and the affordances of contemporary computer systems that might actually execute the algorithms, but they essentially remain abstract solutions to abstract problems. This quality requires critical attention to both the ways in which algorithms are historically determined and their evolution and development over time. Algorithms and methods, despite being theoretical, are almost always imagined and developed within the limits of available computational resources and in relation to a practical execution of an imagined solution. What software developers call elegant solutions are frequently deemed elegant because of the handling of these limitations.[24] What we might call a rigid or overly rapid historicization would examine either the moment of creation or specific implementations of algorithms.[25] This approach fails to take into account the influence of prior solutions and methods on the development of algorithms and changes introduced into the algorithm through the adapted evolution of the solution to the new material conditions found within the altered problem space.

Despite the received wisdom and common understanding of modernity as essentially a narrative of progress, with the new positioned as or seeming to appear on an ever-receding horizon that succeeds and obsoletes the present, the history of ideas, and especially the history of ideas in technology, is never linear. Any account of the design, operation, and use of any contemporary technology is incomplete without situating a particular technological solution within the history of other attempts to solve the same or similar problems. Solutions once considered unfeasible become possible with changed material conditions. This is as true of improvements in processing speeds and storage capacity as it is with social factors, including comfort with particular interfaces and with the directives and priorities of government and/or corporate entities. Approaches to technical problem solving, to use the example of the long-held desire for automatic object recognition from visual stimuli, did not simply appear at a particular point in time as the compiled results of solved lower-level processes. Rather, there have been a whole host of solutions, many of which satisfied the framing conditions of the original problem, alongside both higher- and lower-level algorithms and procedures. The nature of these technological developments requires nuanced understandings of the historical conditions, what was feasible within these conditions, and what was imagined in excess of the possible.

Why examine these algorithms and the history of the field of computer vision? What makes computer vision worthy of critical analysis? Complex and computationally intensive computer vision techniques are highly visible within the public perception of artificial intelligence. At the same time, these techniques have become embedded within everyday life, from the smartphones in our pockets to our cars. We encounter the products of computer vision in use when we drive through highway tollbooths and when we stop at red lights. These omnipresent algorithms capture and analyze our appearance and our movements. Amazon's development and marketing of cloud-based facial recognition technology as a law enforcement tool should give us pause for many reasons, including its unsubstantiated claims of efficacy, its presence in the history of privatization of public services, and, most importantly, as theorists like Simone Browne argue, for the connections between this highly biased mode of policing and the ongoing criminalization of Black bodies.[26] Computer vision technologies have been a crucial part of the ongoing reconfiguration of the senses, especially in relation to perception, during the past fifty years.

Computer vision approaches are also an ideal site for examining the historical development of algorithms because, as this book will demonstrate, the algorithms and approaches deployed within computer vision are similar to, or sometimes exactly the same as, those used in other computational applications, including many common procedures in text mining. For many applications, data are data, whether we are considering an encoded continuous string of characters or massive matrices of image data; some of the basic approaches to make meaning from these already meaningful representations are quite simple. Computer vision has often been located as a subfield of artificial intelligence despite its use of many trivial (that is, not computationally intensive) algorithms. Computer vision methods emerged from many of the same research labs and programs and were supported by many of the same government funding agencies as those involved in the creation of machine learning and artificial intelligence algorithms.[27] As Richard O. Duda and Peter E. Hart, authors of an important and foundational textbook on pattern-recognition algorithms, explain, "Attempts to find domain-independent procedures for constructing these vector representations have not yielded generally useful results. Instead, every problem area has acquired a collection of procedures suited to its special characteristics. Of the many areas of interest, the pictorial domain has received by far the most attention. Furthermore, work in this area has progressed from picture classification to picture analysis and description."[28] Many early computer vision projects explicitly combined the objectives of image recognition or understanding with those of automatic classification. The history of computer vision crosses a boundary within artificial intelligence when the field recoiled from the failed promises of automated approaches to the development of what were called expert systems—rule-based complex systems, like those found in sites like hospitals, that encoded the knowledge of specialists into hierarchical decision trees that helped generalists diagnose problems, and that worked alongside and integrated statistical approaches but depended on long rules that defined responses to specific known conditions—and back to automated decision-making systems with the increase in available data in terms of size and resolution, improved algorithms, and greater computational and storage capabilities. When computer vision techniques ran into technical limitations in analyzing visual representations of the world, researchers turned to the encoding of knowledge about the world from expert humans as a supplement to visual data. Artificial intelligence's turn to expert systems was made possible by encoding knowledge that could not easily be learned from directly digitizing complex objects like images. Via this encounter, computer vision became for many a subfield of a reimagined artificial intelligence that was especially invested in making different forms of knowledge comparable and computable.

The lack of mobility of computing during the invention of computer vision and the limited power of portable computers until the early twenty-first century curbed the imagination of computer perception and understanding. The perceptual space for almost all computer applications since their inception has been virtual. As a result of the high degree of modularity in computing, the discrete logic that separates input devices, sensors, and CPUs, the space in which the computing of perception takes place cannot be easily mapped. In the 1970s, Duda and Hart were interested in "the ability of machines to perceive their environment."[29] While they were working with semimobile computers and sensors, mostly in the form of an early semiautonomous robot named Shakey, for most programmers, the notion of a computer's environment was relatively restricted to secure-access, climate-controlled rooms. Even Shakey was closely linked to very large, immobile computing systems that performed almost all the complicated computations required to permit Shakey to have any mobility whatsoever. The narrative of the increasing portability of computation, marked by the shrinking of computers, cameras, and other sensors, would have us believe that computer perception has become less virtual over time as the activity of sensing the environment increasingly takes place in the same device as the computation. Yet what this narrative gets wrong is the degree to which all computer vision depends on other imaginations of the sensed space and scene, almost all derived not from other computationally sensed descriptions of the environment but from what we might call human sense: the names and categorization of visual objects, knowledge of how these objects are constructed, understanding of how they move throughout the world, and a sense of which objects might be meaningful and which might not.

Reading Computer Vision

It is not exactly clear what might form the proper object of study for algorithms. Examining the systems or devices in which they are used might not provide a broad enough frame for understanding their development and operation. We might want to examine its design specifications, but this will not tell us how the algorithm works in practice. Do we attempt to examine it as implemented in code? Do we examine the original or reference implementation of this algorithm and code? If the algorithm has had a long life, might we prefer its implementation in a present application and technological system? In attempting to answer these questions, it quickly becomes apparent that some way of understanding the historicity of algorithms is needed. Algorithms, unlike programs, are not static. An algorithm may become fixed as implemented in a particular programming language or compiled into a specific program, but the algorithm itself is not the implementation. Algorithms undergo transformations over time. These transformations are not simply additive; they cannot be understood as complete replacements of prior, now obsolete algorithms. Examining them as a class of similar methods is not always useful. Algorithms are organized by computer scientists into families, and while those so grouped might share a resemblance, sometimes they are used for wildly different tasks. The same computer vision algorithms and methods used at one point for facial recognition were also used for topographical segmentation of aerial photographs. Examining a class of sorting algorithms might not tell us as much about the lines of influence as beginning with one particular algorithm.

We might productively think of the history of algorithms as a history of remediation. "Remediation" is a term from media studies used by Jay David Bolter and Richard Gruisin to describe how new media frequently promise to reform or refashion prior media; in the process, this promise "inevitably leads us to become aware of the new medium as a medium," thus creating what they refer to as hypermediacy.[30] Remediation, in the case of the abstractions we are calling algorithms, might involve the reimplementation of an algorithm from one language to another; in the process, the algorithm may prove more or less successful or popular than prior implementations. Reimplementation might involve simply a translation of operations, termed "porting" by computer engineers, or an utter reimaging of parts of the algorithm in moving from a computer architecture with a single CPU to a multiprocessing or parallel architecture. Another form of algorithmic remediation might involve bundling a set of algorithms together with shared and optimized libraries. No longer single purpose, components of one algorithm may now be used by others. The incorporation of some algorithms into popular tool kits or applications may also function to give greater prominence to a particular algorithm. This might have less to do with the attributes of a particular method than the ease with which it might work alongside others included in a particular package.

Cultural studies offers a crucial framework that lets us understand both historical and present-day implementations of algorithms, the transformation of algorithms over time, and the complex temporality of computer vision. Algorithms, of course, are expressions of culture; they are increasingly necessary to understanding the functioning of contemporary culture, including sorting and prioritizing existing information and content and producing new cultural objects.[31] When I say algorithms are cultural, I mean that they are generated and used within a complex social scene that exceeds their technical function. Humanists and social scientists have been studying the culture of algorithmic production. They are interested in how people who interact with algorithms describe them, and they use these people, both developers and users, as sources of knowledge about algorithms. Nick Seaver, one such social scientist, argues that an ethnographic approach is the best way to study algorithms because they are essentially unstable artifacts; contrary to the common perception that algorithms are intended primarily for machine consumption, they are best understood as enacted through multiple different interactions with interested actors.[32] Yanni Alexander Loukissas posits that the historical split between the algorithm and the data on which it operates, as used by interpreters of algorithms, is not supportable because these users are connected to that which they seek to model: "Algorithms are local, not in small part because they rely on data for their development and testing. I would take that argument one step further: collections of data and algorithms should not be considered as entirely independent components of computation. Indeed, they are entangled with each one another, materially and historically."[33] This entanglement of data and algorithm takes the form of a network of unsettled relations, but it could also be described in terms of a feedback loop, in which each modifies the other, or recursion, in which the output becomes input for the same function, called by itself. Loukissas's critique is invested in localizing algorithms, which is to say reading them in relation to their material and historical conditions, and understanding how these algorithms modify that which they seek to model. He gives the example of natural language processing, which has historically been trained on data derived from news reporting, used in the analysis and production of text-based news data. In the machine learning context in which most computer vision algorithms are located, the difference between training and testing, model creation and validation, and training and application demonstrate such an entangled relation. This introduces a minimal historicity into the model via the temporal gap between data collected, refined, and evaluated for training and development purposes and those data acquired in the wild. Yet it is not sufficient to examine only the training data to understand the input supplied by operators to the model as the only possible location for bias and for politics. The critical analysis of complex computer systems using machine learning or artificial intelligence must also give serious attention to the cultural and intellectual historicity of these algorithms.

In the case of computer vision algorithms, the stakes are high. Reality increasingly becomes augmented by the products of computer vision, which themselves are increasingly models of that now augmented reality. The genealogy and historicity of these algorithms matter because the models of the world that they impose back on that world through projection, overlays, and manipulation bring with them the biases and assumptions of other historical moments. Prior norms and hierarchies of knowledge are wrapped up in these models. An example of how cultural norms become embedded in the operation of technology and algorithmic thinking can be found in the development of Norbert Weiner's cybernetics. In Peter Galison's historicization of cybernetic theory, Galison argues that Weiner's conceptualization of control mechanisms were organized around what he called a Manichean struggle with an oppositional intelligence. This struggle was deeply indebted to Weiner's experiences as a mathematician during World War II. "In general," Galison argues, "the cultural meaning of concepts or practices . . . is indissolubly tied to their genealogy. To understand the specific cultural meaning of the cybernetic devices is necessarily to track them back to the wartime vision of the pilot-as-servomechanism."[34] Cybernetic technologies were thus subject to one of the overriding cultural logics of the period—what Galison refers to as the ontology of the enemy. Computer vision's initial ontologies were created alongside, and sometimes coarticulated with, cybernetics by wartime scientists and Cold War researchers. Research in computer vision thus joined with these field-adjacent discoveries in cognitive science and cybernetics to produce reconfigurations in the discourse of vision and perception that, as Orit Halpern argues, "linked the nascent neurosciences of the period to broader changes in governmentality relating to how perception, cognition, and power were organized."[35] The logics and norms of these early moments structure the affordances and limitations of these perception-capturing and -altering technologies. In the case of pattern recognition operating on a set of sample template images, the normative assumptions active in the collection of these samples from one moment will eventually be applied to another, unless the sample templates are continually updated or are replaced wholesale with normative assumptions of the present. This sort of historicity accretes in both data and model. Another type of algorithmic historicity is to be found within the algorithm itself. Some algorithms are unable to handle the necessary scale in terms of input values or features, or number of classes or categories needed to model phenomena or objects, as our understandings of these change and become more complex. Computer vision algorithms and procedures, as well as many others used for similar tasks in other representational domains, thus have a complex temporality that is at times entangled, lagging, and out of step with the world in which they are used.

The major algorithms and techniques examined in the following chapters—including the Perceptron, pattern matching, pictorial structures, and the Hough transform—were created during the infancy of computer vision, but all remain with us today, although in slightly altered forms.[36] The algorithms and methods found in this book were selected because they were present during the development of the field and have persisted to our contemporary moment, although sometimes, in the case of a remediation from one computer environment to another, they take a dramatically different form than the one found in their initial design. In order to perceive the cultural import of these algorithms in the present, we need to understand the conditions through which they were imaginatively called into being and first implemented in software and hardware; we must also address their historical development since this moment. This book thus takes a genealogical approach to these particular cultural objects and to critical algorithm studies more generally. Genealogy, as Michel Foucault writes, "is gray, meticulous, and patiently documentary. It operates on a field of entangled and confused parchments, on documents that have been scratched over and recopied many times."[37] Genealogical analysis has been applied to many objects of interest to humanists, but the archive examined has primarily been, as Foucault suggests, textual and material. In examining technology through a genealogical lens, we can read textual artifacts connected with the creation of devices and code: the project proposals and annual reports addressed to funding agencies, the scientific articles in which these inventions are described for other researchers, patent applications in which the inventors make their claims for the originality of their ideas, and public reporting on the devices, such as press releases and news stories.

The computer vision research conducted and the applications to which this research was applied during the decade between 1960 and 1970 thus form the archive animating the following chapters. The history of computer vision requires understanding these ideas and technologies within their own context as well as their continued use and ongoing development. Algorithms, as I have argued, are abstractions. They are inscribed answers to (usually) well-defined problems. They are reduced sets of instructions and compressed models of the world. They are imagined and shaped in response to certain material conditions. Yet algorithms are also alive. They are discursive objects in which we see researchers and computer scientists think and carry on the work of others. They contain within them traces of the past—both as material records of particular implementations of algorithms and in their abstract form. Each chapter takes up algorithms, code, and projects developed during this period, but the primary archive is the discursive field in which these ideas were imagined, designed, articulated, tested, and debated.

Chapter 1 provides an overview and history of the field of research known as computer vision and the application, developments, and activities arising from the computational manipulation of digital images. By focusing on the representational rather than simulative aspects of manipulating digital images, what earlier researchers took to be the founding difference between computer graphics and computer vision, this chapter connects the various threads of psychological, physiological, and computational research that were woven together in the project to develop the first pattern-recognition systems. This chapter also introduces a methodology for more historically orientated critical algorithm studies and makes the argument that cultural studies methodologies can best help us understand the ways in which the historical design paradigms of algorithms and the affordances of early implementations have become congealed within the ongoing revision and contemporary use of these algorithms.

Chapter 2 situates the previously murky origin story of the development of a major class of computer vision techniques within the field of artificial intelligence through an examination of Frank Rosenblatt's Perceptron, an algorithm as well as a physical device. Rosenblatt, a research psychologist trained and employed by Cornell University, developed the Perceptron with the financial support of the U.S. Navy for Cornell Aeronautical Laboratory in Buffalo, New York, at the time a research institution affiliated with Cornell University. The Perceptron took the form of a specialized computing machine, a digital brain model and an alternative to general-purpose digital computers, that was initially constructed in 1957. Rosenblatt's Perceptron was an early neural network—a nerve network, as Rosenblatt called it at the time—that in its initial form was capable of producing a linear classification between two categories. The algorithm was prematurely introduced to the public by the U.S. military and was received with much excitement by the press. The high expectations produced by its public demonstration and the myriad dreamed-up uses for this new computing paradigm were well beyond the capabilities of 1950s-era computing hardware. Rosenblatt's device failed to live up to these expectations.

It was not a robust system and was nothing like a general artificial intelligence, yet the theoretical concepts and the algorithm itself laid the foundations for contemporary machine learning. The Perceptron's limitations and the resulting debate over this neural computer within the nascent computer science community were partially responsible for the decades of reduction in funding for machine learning and neural network technology that followed. This chapter takes up the discourse and design of Rosenblatt's Perceptron and several other algorithms that were also implemented in specialized hardware, as well as simulated on general-purpose digital computers, that were collectively referred to as learning machines. These machines appeared alongside the development of more advanced digital computers and the field of computer science.

Chapter 3 analyzes a crucial moment of transformation in the field of computer vision: the movement from what was called automatic photointerpretation to image understanding and finally computer vision proper. This transformation happened almost entirely within a body of research dedicated to military applications, initially for operations in the Vietnam War and then for ongoing Cold War surveillance and targeting applications. The image and object description methods that were developed at this moment continue to be used in many popular face-recognition algorithms, and the mapping of facial features owes much to the solutions found for the mapping of topological features in aerial photography collected during the Vietnam War.

Chapter 4 examines a particular class of low-level computer vision techniques that were brought into being through the Shakey Project, a 1960s-era DARPA-funded research project into computer-driven robotic automatons. These automatons, of which Shakey was the prototype, were initially expected by their funders to be able to be deployed by the U.S. military as reconnaissance devices in hostile climates and used in support of the ongoing war in Vietnam. Shakey altered the ontology of computer vision by directly connecting the sensor and the site of computation, making mobile the model of space reconstructed by computer vision algorithms. This required developing reliable methods of tracking object boundaries as the image acquisition site moved through space. The Shakey prototype robot was developed at Stanford Research Institute/SRI International, where it was used as a platform to advance several different research areas connected to the related fields of artificial intelligence and computer vision. While the Shakey project never resulted in the production of a military-grade automaton, and several of the researchers working on Shakey left the project after changes in DARPA requirements that necessitated a military application for all funded projects, the original military goals lingered on in the research projects and algorithms that came out of this project.

The Coda offers a short, critical reading of the OpenCV computer vision tool kit. This tool kit provides reference implementations of a number of important algorithms, many of which are introduced in the previous chapters, that were either created during the early moments of the field or are the direct descendants of these initial algorithms. Rather than interpreting the use of computer vision applications developed with these packages, the Coda positions this tool kit as comprising the core enabling technologies of seeing for the contemporary moment. In closing, the Coda offers an account of a special class of images created by recent computer vision models, including generative adversarial networks and convolutional neural networks that visualize aspects of hidden layers, weights, and coefficients used for image classification decisions. Reading these images as a version of what filmmaker and theorist Harun Farocki has called an operative image, an artifact of the intensifying machine-driven sightless seeing of the late twentieth and early twenty-first centuries, enables us to understand their place in the genealogy of computer vision addressed in the previous chapters.

« 1 »

Computer Vision

> If we want to give our computers eyes, we must first give them an education in the facts of life.
>
> —Azriel Rosenfeld, *Picture Processing by Computer*

> Perception is not a science of the world, it is not even an act, a deliberate taking up of a position; it is the background from which all acts stand out, and is presupposed by them.
>
> —Maurice Merleau-Ponty, *Phenomenology of Perception*

Aleksandr Georgievich Arkadev and Emmanuil Markovich Braverman, two Russian engineers, introduce their 1967 coauthored book *Computers and Pattern Recognition* by making the case that the mechanics behind two now quite familiar computer vision activities are worth studying and solving: "The problem of 'how do we distinguish male portraits from female ones,' or letters 'a' from 'b,' i.e., the problem of image recognition, is much more interesting and complicated than might at first appear."[1] Prescient and provoking in their presentation of computer vision problems, Arkadev and Braverman's short Cold War–era text covers a range of recently developed machine learning methods used for computer vision, including those developed in both the United States and Russia. The binary model of decision-making, a model in which one forces an algorithmically determined decision between the categories "male" and "female" or between representations of the characters "a" and "b," may not have been entirely a product of a Cold War culture organized around an us-versus-them dichotomy, but it was certainly compatible with the dominant imperatives of the time.[2] Dividing the world into binary and separable regions was both policy and best practice in computer research at the time. Arkadev and Braverman write through the international scholarship and research networks that managed to cross the Iron Curtain and use their survey to build bridges between American and Soviet work on advanced methods for analyzing digital images. Braverman conducted research at the Institute of Control Problems (Institut Problem Upravleniya) at the USSR National Academy of Sciences in Moscow, a center with significant governmental support dedicated to applying cybernetic theories to societal problems.[3] Much research was conducted in the Soviet Union on computer vision and pattern recognition during this period and later. In the late 1980s, U.S. researchers would assess these efforts as strong and comparable to the theoretical work in computer vision in the United States but noted a lack of access to recent hardware and software.[4] While Arkadev and Braverman's contribution demonstrates the global interest in computer vision—Italian and British researchers were also present at important early conferences and developed their own software and hardware for computer vision—the direction and development of computer vision during its early years was driven predominately by a small number of university labs funded by U.S. federal agencies with military connections, most importantly the Defense Advanced Research Projects Agency (DARPA), an organization created in 1958 as ARPA by President Dwight Eisenhower in response to the USSR's demonstrations of their technological achievements through the Sputnik program.[5]

The changing configurations of mid-twentieth-century visual culture made computer vision both imaginable and possible. The laboratory environment for computer vision experiments was modeled on the studio environment, complete with soundproof control rooms and tape storage.[6] The technological apparatus of early computer vision also borrowed heavily from the television studio. Commercially available television cameras were directed toward photographic images to acquire image data. These cameras were also eventually mounted on mobile devices to produce a movable frame. The scanning mechanism of the television raster provided a model for the storage of digital photographic data. Televisual signal processing provided analogs and at times solutions for several problems encountered in conceptualizing encoded images and transmitting these as digital data. The rapid proliferation of images during the twentieth century, especially those created and collected by U.S. military and intelligence organizations, produced a mass of images that required the assistance of machine processing if these images were to become a source of information. The historical formation of computer vision, its material conditions and configurations, and the priorities that shaped the initial research in the field are important to understanding these now ubiquitous technologies. In becoming acquainted with the early history of computer vision, we will simultaneously see both some of the remarkable successes of these new technologies and the degree to which some of the major problems present at the founding moment of the field remained unsolved.

The umbrella term "computer vision" now serves as an organizational rubric for applications and algorithms involved in both high- and low-level image processing. This was not always the case. Computer vision was accepted as the name of an area of research and practice involving the manipulation of digital images and the recognition of objects represented within these images only after a series of disappointing developments and encounters with the research products of other fields. Now ubiquitous in almost all aspects of contemporary life, computer vision has become something of a computational success story. This research, however, can only be framed as a success retrospectively, and only though narratives that have undergone necessary reconfigurations in response to changing public perceptions and the founding goals that were established primarily by military and government funding agencies. These early efforts attempted to use relatively simple pattern-matching techniques as a substitute for more complex human perception tasks. Unable to deliver on the promises of automated perception made to these military funders, researchers reframed aspects of their projects and shifted to examining other sources of knowledge to address the growing archives of photographic images. After the publication of well-known critiques by other computer scientists, some of the major algorithms that appeared at the birth of computer vision were also shelved as impractical or not able to handle the complexity. The term "computer vision" refers not only to this field of research but also to the practical application of these methods. The discourse of computer vision depends on a redefinition of vision and seeing as the neutral extraction of information from data. This information is presented as superior to a now suspect human vision. The technology industry has thus deployed the descriptive label "computer vision" to signal advanced tools and methods that it presents as augmenting or replacing the visual field.[7] It is the goal of this chapter to excavate and historicize the theory of vision underwriting this discourse and the important political and ethical concerns raised by this way of perceiving the world.

Representation and Simulation

Computer vision as a field originated with a group of researchers who had situated themselves in opposition to another emerging field, computer graphics. The difference between these two activities, as they understood it, rested primarily on computer vision's emphasis on analyzing rather than creating digital images. These are both fields of research and practices, and in the case of computer technology, field priorities and imperatives became embedded within the material objects and software that eventually landed in the hands of users and consumers. It is important to understand the distinct articulations of computer graphics and computer vision, at least as they existed during the founding moments of this field, because the assumptions and authorizing theories about these fields' objects inform the creation and analysis of these very objects. The objects in this case are digital. Yuk Hui, a philosopher of technology, in *On the Existence of Digital Objects,* defines them as those "objects that take shape on a screen or hide in the back end of a computer program, composed of data and metadata regulated by structures or schemas."[8] Hui positions digital objects as a new genre or class of object, joining natural and technical objects, in order to produce an ontological theory that can help us understand the existence of these objects, and in particular their genesis through a "process of concretization and materialization, first of forms, second of explicit relations and connections between objects."[9] Hui's understanding of concretization, a concept he borrows from Gilbert Simondon's account of technical objects, concerns what he calls, following Simondon, the recurrent causality of digital objects. For Hui, this causality necessitates examining digital objects in relation to their creation or genesis within their specific digital milieu—that is, the environment in which these objects are created, assembled, and used. While both computer vision and computer graphics deal with the creation and manipulation of digital objects that concern visuality, it was the creation of these objects within different recurrent causalities that defined the boundaries of these fields for early researchers.

Azriel Rosenfeld, a computer scientist employed by the University of Maryland, College Park, took on the role as one of the primary organizers of much of the early technical history of computer vision. In his 1973 state-of-the-field survey, Rosenfeld explains that he is concerned "only with the computer processing of given pictures, and not with pictures that have been synthesized by the computer; this is the fundamental distinction between picture processing and *computer graphics*."[10] For Rosenfeld, digitized images were not synthetic but representational. While both are to be understood as digital objects in the sense that they are created and concretized by algorithms within a milieu of various encoding schemes and standards, for Rosenfeld, digitized image data, even if modified algorithmically after digitization, needed to be causally linked back to sensed image data rather than a machine-approximated synthetic image. While Lev Manovich argues that computer graphics and computer vision "were born simultaneously," his emphasis on the need for computer vision algorithms to understand images in three dimensions obscures the early two-dimensional problem that had organized the field: the analysis of top-down aerial surveillance imagery for photointepretation.[11] A model of three-dimensional space was not required—and is still not necessary—for many computer vision tasks, including machine learning–based object identification. The digital objects used and created by computer vision do not necessarily require the sort of perspectival modeling necessary for computer graphics. Another way of understanding the difference in terms of the causal origins of these digital objects can be found in Jacob Gaboury's history of computer graphics, *Image Objects: An Archeology of Computer Graphics*. Gaboury demonstrates the importance of a "distinct theory of computer graphics as simultaneously screen image and simulated object."[12] He gives the name "simulation" to what Rosenfeld termed "synthesis" and explains that the objects in computer graphics have no existence outside of the computing environment. There is also a crucial link between the screen image and the digital visual object that provides another key differentiator for computer vision, as these methods and algorithms were initially developed without a capacity or even need for display. The visualization of digital images or transformations applied to these images was not important to early computer vision researchers. What was important was the categorization or identification of objects within sensed image data that, with as much certainty as possible, could be linked back to the real.

Rosenfeld, while not exactly anticipating ontological accounts of digital images like the one proposed by Yuk Hui, works to clarify that while picture data are "mathematical objects," they are not objects in and for themselves but rather are representationally linked to real objects: "What makes picture process a subject in its own right is that it deals with pictures that are not merely arbitrary functions or matrices, but that are pictures *of* something—which purport to represent a real scene (terrain, microscope slide, . . .) or an ideal symbol (such as an alphanumeric character)."[13] Rosenfeld makes this argument early—*Picture Processing by Computer* was published in 1969—before the consolidation of his field of research into computer vision. This representational aspect of the images used in computer vision presents some complications for how we think about the manipulation of image data through various algorithms and methods. When algorithms operate on second-order representations—that is, a numerical representation of data that is understood to represent sensed scenes within reality—they are functioning on abstract structures and numerical values that have an increasingly distant isomorphic relation to the objects that they are assumed to be representing. The representational status of these abstract structures and features will need to be examined in relation to particular algorithms and methods. This particular boundary, the distinction between simulated and representational digital objects, that has organized computer vision and computer graphics has become, in the twenty-first century, increasingly less certain as computer vision methods are iteratively applied to what Jacob Gaboury calls simulated objects. The production of images or image-like data within contemporary neural network approaches also frequently involves the creation of images that have either no relation or only the most minor connection to Rosenfeld's sense of "real" scenes and are much more conceptually similar to synthetic images. Other examples of this sort of increasingly boundary-crossing creative/analytical use of computer vision methods would be the image filters or automatic corrections applied to digital images within a camera or through an application such as Instagram, or the creation of "synthetic" images with generative adversarial networks or other deep-learning techniques.[14] At the same time, the increased interest in automation of higher-speed robotics and self-driving vehicles in the contemporary moment has intensified computer vision's need to focus on problems connected to the extraction of information from real-time representations of the real world.

The Ontology of the Digital Image

While the encoding of visual information through sampling and quantization produces a new representation of visual scenes, the primary concern in understanding the ontological status of the image within computer vision is the altered sense of space produced by the entire social and technical apparatus that has made these methods possible. The source objects of computer vision are not the world but rather a representation of it. The desire first articulated in the 1950s and 1960s for high-quality "live" images or video streams has only recently become a possibility. During the development of the algorithms analyzed in this book, photographs were most commonly taken with what was then high-resolution black-and-white 70mm film and then "sensed" or digitized by expensive and frequently specialized equipment. Photography and computer vision are linked by necessity because computer vision depends on the prior existence of photographic images, but there is also a changing relation and history between perception and the image that connects these two ways of seeing. Computer vision makes an important contribution to the objective promise of photography by extending the autonomy of the image through mechanization, but it also introduces other aspects of more subtle mediation through obscured selection criteria, framings, labeling, and other human activities that trouble the singular genesis of the image. These mediations function as what might be conceived of as hidden layers, to borrow an apt phrase from the language of neural networks.

The assumed objectivity of the photographic image and its digital surrogate, as invoked in the discussion above, is the key difference between computer graphics and computer vision. This representational aspect of image data within computer vision, by virtue of a causal chain of transformations leading back to the sensed image of reality, underwrites the authority asserted by demonstrations of an algorithm's ability to discriminate and classify data. These are not classifications of artificial or simulated objects but rather representations of real objects in the world. In this self-understanding, computer vision attaches itself to long-standing claims associated with photographic realism, in terms of both a dependency on the status of the photographic image as such, without which it would not function in its early development, and as the supporting theory for identifying the computationally manipulated digital image with what it has been understood to represent. André Bazin's "The Ontology of the Photographic Image" provides a classic definition of photographic realism: "The photographic image is the object itself, the object freed from the conditions of time and space that govern it."[15] This is to say that the image of an object is equivalent to the object itself. Bazin's investment in photographic realism is shared by later theorists who want to understand the power and appeal of photographic realism, although some, like Susan Sontag, are willing to grant them power beyond just their status as documentary evidence. In *On Photography,* Sontag describes photographic images as found objects in a way that would be compelling to computer vision researchers: "Photographs are, of course, artifacts. But their appeal is that they also seem, in a world littered with photographic relics, to have the status of found objects—unpremeditated slices of the world. Thus, they trade simultaneously on the prestige of art and the magic of the real. They are clouds of fantasy and pellets of information."[16] It is precisely this extraction of "pellets of information" from images that computer vision seeks to automatize. The challenge would be the minimization of these other aspects of photographs, what Sontag terms "clouds of fantasy." For Bazin, the more mechanical the photographic process, the better:

> The personality of the photographer enters into the proceedings only in his selection of the object to be photographed and by way of the purpose he has in mind. Although the final result may reflect something of his personality, this does not play the same role as is played by that of the painter. All the arts are based on the presence of man, only photography derives an advantage from his absence.[17]

The degree to which the photographer as a figure for the perceptual subject is absent from the pattern-recognition process has become extremely important to arguments about computer vision, and not just in supporting its claims to photographic realism on behalf of image data. Computer vision depends on the documentary status granted to these digital artifacts, even or perhaps especially when the origins of the "found object" is not a photographic image but the virtual artifact of sensed image data, to support its claim to provide unmediated access to the real world.[18]

Computer vision certainly could be said to prefer the clean, straight lines of "objective" documentary images. The less natural-appearing an image is, the better suited it is for the purposes of computer vision. This is to say that computer vision methods designed to separate and identify individual objects are best matched with those images with a lack of visual admixture—those that represent an already cleanly separated and segmented world. Computer vision, like its enabling statistical algorithms, assumes the eventual separability of the world. High-resolution and cropped images that are centered on the object of interest are preferred for feature extraction because the features to be extracted, be they grids of pixels or higher-level features drawn from already known patterns, will have a higher probability of representing the depicted object. Computer vision knows no difference between subject and object. These techniques have been designed for a prototypical constructed scene: a visual representation of a series of objects created and placed in the scene by humans. This preference for highly structured and simplified scenes is not just the result of training images composed of advertising-like object-focused and -oriented images but is built into the foundations of computer vision. This is because these applications have been trained to read not images of the unruly natural world but rather scenes understood to have encoded signifiers of specifically human activity—the organization of the world through lines demarcating distinct spaces and objects.

The use of historical photographic images in computer vision only adds to the complexity of the interpretive process for examining the results of algorithmic classification and object identification. We should ask familiar historiographic questions about the image itself, the positioning of the photographer vis-à-vis the subject of the photograph, and the preservation and archival processes that resulted in this image's survival. This is to ask, among many other questions, how representative is this image? If photography had been imagined as primarily a documentary technology, then computer vision raises the stakes of this objectivity through scale. Rather than requiring an interpreter or observer of live video feeds, for example, computer vision algorithms can scan every frame of video to find unusual activity or known figures and shapes—a mind-numbing task for humans. Video, when analyzed frame by frame, is a rich source of image data, as objects will most likely be viewed from several different angles as they move within the frame or as the camera itself moves. Likewise, the presence of partially occluded objects can also increase the number of samples for matching future partial representations of these objects.

Computer vision carries with it something that might be imaginatively conceptualized as the afterimage of the photographic image and the photographer. This is because there is a phantom photographer behind every instance of computer vision. This phantom photographer holds tacit knowledge about the world and what is potentially meaningful. It is a mediating figure between the probabilities produced with statistical methods and the always already displaced human perceivers. This phantom photographer is not a singular subject but multiple; it might be thought of as an amalgam or an average of the positions of all the photographers/perceiving subjects found within images comprising the training data set. The discomfort with this displacement of the photographer is perhaps one reason why so many computer artists have made such heavy use of image averaging. It is not the subject of the average as much as their own displaced, shifting, and amorphous position that they seek to explore by averaging a large number of photographic images. When an artist like Jason Salavon, for example, averages digitized historical magazine cover images, we might read his averaged image as demonstrating an anxiety about the positioning of artists such as himself in relation to the archive more than an account of the archive as such.[19] The choices of these prior photographers belong, of course, to different historical moments, and each is particular. Computer vision assumes that these differences will be erased, as each image is understood to contribute to a triangulation of the photographed object—to a slightly blurred generality rather than a focused specificity. Yet traces of these particular positions in relation to the object are preserved in the image data, and the choices of objects and included variation define the possibilities of any generality.[20]

We might think through some of the ontological problems invoked by computer vision through Paul Virilio's understanding of sightless vision, a direct sensing of the environment by machines. While Virilio introduces this problematic of sightless vision in terms that suggest it marks the advent of the epoch of technological development in which "the video camera would be controlled by computer," the implications of his analysis of the "vision machine" are more ontological in nature.[21] He notes a transformation made possible by computer vision that relocates the site of observation and reconnaissance from human observers to the environment itself. This raises, as he reminds us, "the philosophical question of the *splitting of viewpoint,* the sharing of perception of the environment between the animate (the living subject) and the inanimate (the object, the seeing machine)."[22] It is not the loss of an objective human vision that bothers Virilio, for human vision is indeed quite animated, as physiological research into eye movement and retinal retention has destabilized any account of objective vision, but rather the presence of perception devices in the perceived environment, especially when that environment includes humans. His invocation of the split viewpoint is productive for thinking about one particular positioning of computer vision that would seek to displace the subject entirely from the scene of perception. This idea, in fact, is found in many accounts of the operation of automated recognition activities: cameras capture images of sensed scenes, including all visual objects within the field of vision, and are segmented and classified by decision-making algorithms in isolation from biased humans. Yet that machine-dominated scene, which appears stripped of all activities of human visual perception, is only made possible through a scene split in which we have another scene that is saturated with subjects preperceiving those objects seen by the machine. Sightless seeing is possible insofar as computer vision positions itself before human vision, and only through the displacement of the perceiving subject imperceptibly far ahead in the horizon. This perceiving subject, who shares the split viewpoint with computer vision, might be that phantom photographer—an operator providing feedback based on an individual perception of the environment and the output of the seeing machine, or an evaluative subject who determines to which category a sample image belongs.

Origins of Computer Vision

Early in the 1960s and 1970s, many of the computerized approaches to photographic and object analysis understood themselves as working on pattern recognition, scene analysis, or image understanding. While all three essentially concern the same core problem, the detection of objects represented in photographic images, there were differences in how researchers conceived of digital images and their primary goals. Image analysis was usually positioned in the publications and proposals of researchers as just one mode of pattern recognition; the methods these researchers developed were general-purpose algorithms that they understood to be useful for a number of applications. Two major forces combined to alter the history of computer vision. The first was the growing complexity involved in analyzing digital photographs and the need to develop specific techniques that extract meaningful information from the visual representation of objects. The second involved some early computational models and algorithms created by researchers who were especially interested in using what was understood of biological vision as the basis for their techniques.

Some accounts of the formation of computer vision begin with a short memo that outlined what was thought at the time to be a short-term summer project to implement a computerized visual system. Seymour Papert's Artificial Intelligence Group Memo 100 announcing the "Summer Vision Project" was distributed on July 7, 1966.[23] "This particular task," Papert writes, "was chosen partly because it can be segmented into sub-problems which will allow individuals to work independently and yet participate in the construction of a system complex enough to be a real landmark in the development of 'pattern recognition.'"[24] Other histories of computer vision begin with Lawrence Roberts's 1963 MIT doctoral dissertation addressing the detection of three-dimensional solid figures from digitized photographs.[25] Beginnings are helpful in understanding the development of a field, but the search for origins will frequently produce multiple compelling possibilities, all subject to the present understanding and organization of a field. Many of the aforementioned accounts of computer vision, for example, were written by computer scientists working in those historical moments in which artificial intelligence research had been deemphasized and machine learning methods, including early experiments with neural networks, had been dismissed as impractical. In beginning the history of computer vision with Papert's AI memo or Robert's dissertation, these earlier neural network methods are occluded, as are the projects connected with military reconnaissance that should be understood to motivate much of the early work on pattern recognition. The various entanglements of computer vision and artificial intelligence are key to understanding why and how certain methods were prioritized at different times. Unpacking these histories requires connecting key technological developments to the political, economic, and social contexts that made them possible and that enabled the publication and later adoption of these methods.

By the late 1980s, a number of figures in the field had converged on "computer vision" as the name of their specialized area of computer research. There was general consensus by this point that "computer vision" was the appropriate name for the set of developing computational methods capable of recognizing and manipulating digital images. Azriel Rosenfeld's 1988 essay, "Computer Vision: Basic Principles," introduced this newly reorganized field to the members of the Institute of Electrical and Electronics Engineers (IEEE) in the organization's flagship journal, *Proceedings of the IEEE*. Rosenfield, who by this point had been working on topics in this area of research for more than two decades, glossed "computer vision" in terms that drew on the major research areas and funding priorities that had shaped the field during his professional experience: "*Computer vision* is concerned with extracting information about a scene by analyzing images of that scene. It has many applications, in areas such as document processing, remote sensing, radiology, microscopy, industrial inspection, and robot guidance."[26] Rosenfeld's invocation of scene analysis, a term of art in early research that gestured to the analysis of acquired surveillance images, made it clear that in this defining moment, computer vision understood itself as dedicated to computational analysis of pictorial representations of natural scenes. In short, natural scenes are visual representations of reality. Despite the use of models in the analysis of pictorial representations, computer vision concerns itself with the scene as it exists in the world, not with models or simulations. The methods of computer vision are conceptualized as generating knowledge about these scenes rather than the models themselves.

The term "computer vision" had been circulating almost a decade earlier than Rosenfeld's essay. It was used in the title of Patrick Henry Winston's 1975 edited volume, *The Psychology of Computer Vision,* and in several other collections of essays and books published in the early 1980s.[27] It was also used as the title of Dana H. Ballard and Christopher M. Brown's 1982 textbook on the subject.[28] Ballard and Brown, in a move typical of computer vision researchers, link statistical pattern recognition with image processing, as the former required the latter. They also opened their discussion with a familiar problem to computer vision researchers, and a persistent concern within the field: locating naval ships in an aerial photograph of a busy port. In so doing, Ballard and Brown demonstrate the degree to which, both then and now, surveillance and military reconnaissance tasks motivated computer vision research and shaped the creation and use of core algorithms.

Interdisciplinary Models and Practices

In its rhetoric and practices, the field of computer vision sought to place itself at the intersection of several different areas of research and scientific domains. This field of research spans computer science, neuroscience, and information theory; covers a number of other research areas and disciplines; and has many practical applications. A key example of this rhetorical positioning can be seen in an introductory text authored by Martin A. Fischler and Oscar Firschein, two important computer vision researchers, titled *Intelligence: The Eye, the Brain, and the Computer,* published in 1987. Their book focalizes, in its linking of the question of artificial intelligence to human vision and the project to create computer vision, the major concerns of this book. This entanglement of computer vision with artificial intelligence was essential and not accidental. Fischler and Firschein wrote their book with the full authority of SRI International; their status as employees of this research corporation appears on the title page. *Intelligence* provides a broad overview and was intended to address a general audience, an imagined reader whom they refer to as "an educated layman" and the "*Scientific American* level reader," to recent developments and findings in several fields—in particular the computational work performed at SRI International by Fischler and Firschein and their colleagues. The ambitious scope of *Intelligence*—it frames itself as offering "an intellectual journey into the domain of human and machine intelligence"—demonstrates both the widespread sources and the fields of influence in which these two researchers were involved.[29]

The research context in which computer vision was created—and continues to be developed today—was deeply interdisciplinary. This was an area of research that produced numerical models of biological systems that provided new metaphors for understanding these biological systems in terms of computation. An example can be found in Norbert Wiener's highly influential *Cybernetics, or Control and Communication in the Animal and the Machine* (1948). Wiener's subtitle firmly locates the cybernetics project at the intersection of mechanical and biological systems.[30] Wiener's theories and his explanations of feedback systems, the nature of information, the methods and forms of communication, and the mechanisms of learning toggle between machine and human examples. Research on self-organizing systems and their relation to biological systems was adjacent to, if not overlapping, the work on machine perception. Computer vision and artificial intelligence were, after all, designed to simulate and indeed even sometimes replicate the visual and perceptual systems. Ballard and Brown point out the difficulty of simulating a highly specialized biological system that was not fully understood with generic computing devices: "Computer vision is . . . faced with a very difficult problem; it must reinvent, with general digital hardware, the most basic and yet inaccessible talents of specialized, parallel, and partly analog biological visual systems."[31] The problem, as Arkadev and Braverman point out, is even more complicated than the mere modeling of the as yet unknown, as it was hoped that these simulated perceptual and learning systems would provide researchers with knowledge about the operation of the biological systems that they were based on: "Cognitive machines are also of interest because they simulate a mental process whose mechanism is not yet quite clear; and since simulation of activities of higher nervous systems is one of the recognized methods of investigation of the brain, the development of cognitive machines can also be useful in this direction."[32] Thus, primitive animal models, the more complex human vision systems, and simplified neural models of the brain all served at different times as inspirations for the algorithms and devices that gave rise to machine learning, artificial intelligence, and computer vision. Winston's *The Psychology of Computer Vision* makes a simple case for the importance of computer vision: "Making machines see is an important way to understand how we animals see."[33] Computer vision researchers had lofty goals for some of their projects, and like much of the research into artificial intelligence and learning machines, they held at times conflicting and contradictory senses of what they were doing. Sometimes these early researchers imagined that their newly developed machines and algorithms were modeled on biological vision and perceptual systems; other times, they proposed that these machines were going to help us understand how biological vision worked. We see this same problem in the articulation of artificial intelligence as a model of the brain while simultaneously seeing scientists using the metaphor of the computer to describe the brain's operation.

Fischler and Firschein's effort to explain the workings of human vision provided them with ways to explain the development of computer vision—for example, as the field moved from the simple understanding of the working of a frog's eye that provided the initial version of machine learning metaphor of the receptor—and to imagine computer vision itself as a simulation. For example, in introducing the topic of computational vision, they note, "The computing device should be capable of simulating physical experiments, such as imagining the movement or rearrangement and distortion of objects in the scene to solve a problem or compare the scene with reference scenes stored in memory." They continue, "The computer should have some way of physically interacting with, and sensing, the outside world to build up a database of knowledge and experience."[34] Computer vision has always carried with it this surplus need for knowledge of the world. Vision is sight insofar as sight in this context is the rendering of visual scenes as numerical data, but in computer vision, it is also embedded material and spatial knowledge. This positioning of computer vision makes it dependent on other modes of knowledge extraction and production. Through this adaptative vision, it has for many researchers served to defamiliarize existing accounts of vision as such. Thus, this body of research, from its earliest moments to the present day, has both depended on and influenced psychological and neuroscientific investigations of human vision.[35] Fischler and Firschein's *Intelligence* crystalizes the network connecting artificial intelligence, biological accounts of vision, psychological research into human perception, and computer vision.

Computer Vision Basics

Although this is to some degree a simplification, most digital images are grids or matrices of numerical pixel intensity values. Pixels, or "picture elements," as they were initially named, are simply the smallest units of information about the represented scene. The information contained within a pixel's intensity value generally reflects the amount of light reaching the camera's sensor. These values are the result of sampling and quantization functions that are now built into digital cameras. Digitization, in the form of scanning, provides an analog operation for previously acquired photographs, and several different digitization methods were used in the early days of the development of computer vision. For the sake of simplicity, the following examples are all based on black-and-white or gray-scale images; color images are treated in a similar manner, but where gray-scale images have a single "channel" of values, these images generally have three channels of pixel values: red, green, and blue, or RGB. A typical image is represented by a matrix, which is to say values stored in rows and columns, that contains these pixel values. In this example, the element of the first row is in the topmost left corner, and the last element of the last row is the bottommost right pixel. The size of the pixel—the mapping between the size of the pixel and a real-world measurement such as the pixels-per-inch scale—is usually a value stored alongside an image and is used in displaying the image. Generally, the greater the number of pixels available, the higher the resolution of the image and the greater the verisimilitude to the represented object. The earliest digital images were treated as binary images in which a pixel was either "on" or "off," with the "on" value typically coded as black and given the value 1 and the "off" value coded as white and given the value 0. In 8-bit gray scale, a simple encoding scheme used in picture processing, the pixel intensity values will range from 0 (black) to 255 (white). These 264 values make up the scale of 8-bit gray scale. Such an image represented by a 256 × 256 matrix of 8-bit values will use 512 kilobytes of data. Data compression schemes that function as containers for this image data will modify the actual storage requirements for the image object, but these basic mechanisms of a scale for intensity values combined with a matrix of these values define how digital images are numerically organized.

The above method of encoding sensed visual data, including the notion of a pixel having either "on" or "off" values, was developed alongside early neural networks and growing understandings of the physiological systems related to vision. In explaining this simple image encoding scheme as part of the mathematical and computational "image concept," Arkadev and Braverman make the origins of these grids of pixel values in the growing knowledge of the human perceptual apparatus explicit:

> We consider this way of coding to be natural, because splitting the picture into elements is a basic feature of our own visual apparatus. Indeed, the retina consists of a great many sensitive elements (so called rods and cones) connected by neurons to the visual cortex. The lightsensitive elements of the retina send to the visual cortex, via inter-neurons signals whose intensities depends upon the illumination reaching each individual element. In such a way, pictures projected by the optical system of the eye onto the retina are broken down (via the rods and cones) into fragments and are transmitted element by element, in coded form, to the cortex.[36]

The "off" and "on" of binary pixel values corresponds to the transmission of signals by neurons as well as to the understanding of excitation and inhibition of individual neurons. Early neural networks, such as the Perceptron, conceptualized the network in similar terms as a matrix of values and referred to such matrices, even as they were applied to many different tasks, as the retina. In theory, the coded pixel values are isomorphic representations of projected images. Arkadev and Braverman explain that this understanding, which again is a key feature in the theorization of computer vision, enables computer vision to present itself as addressing vision and not the manipulation of simulated objects: "The one-to-one correspondence between codes and pictures allows us to concentrate on codes only, bearing in mind that a picture can always be reconstructed from its code."[37] In other words, even if our data are quantized and sampled, the sensed or acquired

Figure 1. Image and boundary box around matrix of pixels. Original image Eli Burakian, "Around Campus March 13–14, 2014," 2014. Dartmouth Digital Library Program. Made available by the Trustees of Dartmouth College under a Creative Commons Attribution-NonCommercial license.

representational image exists as the recoverable origin of what may become utterly unrecognizable through a series of algorithmic transformations. Yet it is important to recognize that in the two-dimensional rendering of an image as a matrix of pixel values, the "image" is no longer strictly visual and is not treated by algorithms in any way that resembles seeing.

To understand how this encoded information was proposed to be computationally addressed and manipulated, we can look at the straightforward example of modifying these values. A number of quite simple calculations can be applied to pixel values. To add a border, for example, you can create what is called a binary mask—a new matrix of the same size as the original image matrix with zeroes where you want to place black pixels and ones where you want to preserve the current values—and multiply the values. To make an 8-bit gray-scale image uniformly brighter, you can add a constant value to every pixel; to invert the image, simply subtract the current value from 255. Image rotation is a matter of transposing the matrix of pixel values. To produce an average image from a collection of images, you can average the pixel intensity values. These relatively simple pixel values, accessed as a matrix, provided early computer vision researchers with a "picture function," an approximation of image data through quantization. Encoded visual data are thus available for content-agnostic treatment: the same mathematical or more complicated algorithmic transformation could be used on any image. Early computer vision made heavy use of pixel-oriented manipulations and comparisons to find similar images and for rudimentary detection of known objects within images. The picture function followed the then widely held belief that vision involved the parsing of a field of data rather than interrelations among visual objects.

Objects with a known position, made easier if they exist within a bounded space, can be detected and extracted by addressing only that space from within the matrix. This requires consistent positioning within the frame, but many surveillance tasks concerned with the presence or absence of objects within known spaces—for example, as we will see later, finding airplanes on an airfield or ships in shipyard—can be performed with this basic approach. The bounding of visual space is an identification of pixels within the matrix of values. These submatrices are contiguous pixels, which is to say neighboring in x and y dimensions. Creating these "bounding boxes" around pixels of interest is an extremely common activity in computer vision. Comparing the values of items contained within this bounded region of the matrix of an image understood to be representing this space with and without an object can enable detection of an object. Given two digital images taken at two points in time, we can say that if the absolute value of the difference between the matrices contained by our bounding box exceeds the change in values of those pixels outside the bounding box, then there has likely been some transformation within our area of interest. An airplane now obscures a section of field; the lighter-colored steel of a ship replaces darker-colored water. By moving the bounding box of values a few pixels in each direction, slightly different placements can still be matched. Given samples of several varieties of known objects likely to occupy this region, the type of object can even be detected. Once a region of pixel values has been identified, the masking procedure mentioned above could be used to overlay this extracted object with the contents of another image.

Even in the present moment, this sort of pattern detection and matching remains a core operation of lower-level computer vision.[38] If an image is merely a matrix of pixel values, then patterns

```
[155 156 156 156 154 155 157 159 159 160 161 161 162 163 163]
[156 156 156 157 157 158 158 160 160 160 161 161 162 163 163]
[156 156 158 158 157 158 159 160 160 161 161 161 162 162 163]
[156 157 158 158 158 159 159 160 160 161 161 161 162 162 163]
[157 158 158 159 159 159 159 160 160 161 161 161 162 163 163]
[157 158 158 158 159 159 160 161 161 162 162 162 163 163 163]
[158 158 158 159 159 160 161 161 161 161 162 162 163 164 164]
[158 158 159 160 160 160 161 161 161 161 162 163 164 164 165]
[158 160 160 160 161 161 161 162 161 162 163 163 164 165 164]
[158 160 160 160 161 161 162 162 162 163 163 163 163 164 164]
[158 159 160 160 161 161 163 163 163 164 164 164 164 164 164]
[159 160 160 160 161 162 162 163 163 163 165 164 165 165 165]
[159 160 160 161 162 162 162 163 163 163 165 164 165 166 166]
[160 161 161 161 162 162 163 163 164 164 165 165 166 166 167]
[160 161 162 161 162 162 163 163 164 164 165 165 166 167 167]
[160 161 162 161 162 162 163 163 164 164 165 165 166 167 168]
[161 161 162 162 162 163 163 164 164 164 164 165 166 167 167]
[161 162 163 163 164 164 165 165 164 165 166 166 166 167 168]
[162 162 163 163 164 164 165 165 165 165 166 166 167 167 168]
[163 163 163 163 164 164 165 165 165 165 166 166 167 168 168]
```

Figure 2. Matrix of pixel values included in boundary box in Figure 1.

are nothing more than smaller submatrices of pixel values embedded within the larger image space that can be matched against exact (or, more frequently, similar) sets of values. The simplest procedure using such methods would involve a library of labeled images coupled with an iterative search through a set of target images. Searches might try to match pixel for pixel, or try to match the largest matrix meeting some threshold number of pixels. Such methods, as early computer vision researchers discovered, require images to be composed from a similar enough perspective and angle and to have comparable lighting. Pattern matching with a library of known objects works best with a tightly controlled visual environment, such as the top-down view of highly structured and frequently human-made environments provided by aerial surveillance photography that formed one of the most basic and key target problems for the creation of computer vision.

Reading the Methods of Computer Vision

While a multitude of possible critical approaches could be used for the study of the intricate, multilayered computer systems and algorithms used in computer vision, a historical and cultural analysis that examines the genesis of these methods and their objects is best able to account for their operations and the conditions that made them possible. A driving interest at present is understanding the effects of particular algorithms when used in complex governing and social systems on everyday life—for example, how the mostly opaque algorithms increasingly used in policing encapsulate and reproduce existing biases and assumptions. Algorithms, however, are difficult objects to talk about. As descriptions of solutions to particular problems, algorithms are conceptual and abstract, but when implemented in code and systems, they have a material existence. Like these code implementations, algorithms are imagined in relation to specific conditions—the conditions that framed the problem, what is determined as an important problem to be solved, and the possible solutions from what is thought to be possible at a particular time. The attempt to turn to the design and function of the individual algorithms used within these social systems would require access to these algorithms, many—but not all—of which are trade secrets and proprietary. Add to this the expertise needed to read, frame, and analyze complex code. When code is available, however, bringing to bear the resources of what has come to be known as critical code studies (CCS), an interpretive practice of reading and contextualizing code, can illuminate specific implementations of computer vision algorithms. Even in the absence of code, CCS approaches can examine the output of compiled or inaccessible software to speculate on how the outputted object was created. Yet CCS is not just an explanatory tool for understanding how the algorithm is embedded in code functions. As Mark Marino, a major proponent of this method, puts it, CCS is a humanistic practice that seeks to read culture through code. "A goal of CCS," Marino argues, "is thus not merely to document but to question—to ask what is this object, where did it come from, how did people develop it, how did people use it, and how was it received?"[39]

The code that is part of computer vision works with the discourses founding, surrounding, and permeating the field. Yet CCS would not be sufficient on its own for even the technical understanding of many of the mature algorithms in existence. Because algorithms themselves are frequently reimplemented as a new generation of technologies—technologies including hardware, software, programming paradigms, and languages—understanding changes to the algorithms, not just change in an evolutionary sense but also as a remediation in the terms made available by altered social and technological conditions, requires genealogical analysis.

The prior material forms of algorithms, which is to say the earlier implementations of algorithms, shape the possibilities of later remediations even if there are few or no textual traces remaining within the code of these prior implementations. Algorithms, within computer science, are defined formally. This formal definition enables the demonstration that an algorithm can be proved to be correct—correct meaning it properly executes and terminates. Such proofs are often mathematical in nature and require precise definitions of behavior and input conditions. The precise definitions of algorithms typically are written in pseudo-code notation that is programming language agnostic. The implementation of a particular algorithm in code makes use of the affordances and constraints of the particular language and environment. Algorithm construction is recursive; the formally defined pseudo-code algorithm is written from what is possible with actual computing systems and the algorithm as implemented in a real programming language is justified as correct on the basis of the formal proof of the abstract pseudo code. Although not all algorithms are created through such a rigorous process, when code designed to perform a particular function is remediated in a new computational environment or language, the prior environment extends its influence on the methods of problem solving selected. It is not just earlier formally defined algorithms and implementations of these algorithms that inform the historicity of algorithms but also the existing documentation and test cases, including sample data used to evaluate the algorithm and the original abstract problem for which the algorithm was proposed as a solution.[40]

Computer vision, however, names much more than a related set of image-oriented methods generated out of a field of academic and applied research. Computer vision functions in part to name a collective desire on the part of those eager proponents of surveillance capitalism to automate perception. Computer vision is also a wish to make the world ordered and sensible. It brings focus to our perception through smartphones in the form of face-detection algorithms that focus on our friends and selves. For some, especially the disabled, computer vision may make it easier to operate in the world. In the twenty-first century, it has already made dramatic changes in how we see the world. Augmenting and distorting reality, computer vision, as a technological discourse and as a material apparatus, places itself before the eye. It is now bundled into what Adrian Mackenzie and Anna Munster call "platform seeing," the tighter integration of images, techniques, and devices that organize visual culture in the present.[41] This is a type of vision that is different from human vision, but in efforts to assure us that its knowledge of the world is an objective view and therefore better than human vision, proponents and advocates of computer vision methods elide the historically situated perspective that has enabled these methods to produce the fantasy of a view from nowhere. Computer vision is presented as disinterested; this is why we find it caught up in discussions of policing and monitoring. It is frequently talked about as if it has no capacity for attention, as if it were simultaneously undirected and promiscuous, seeing all and yet focused. Nothing could be further from the truth. Embedded in surveillance systems and in enhancement technologies, computer vision simultaneously observes us and tells us what we are seeing. The camera positioned on a police officer's uniform or in a vehicle is directed, and everyone appearing before its lens has already been captured as a potential subject of computer vision.

The different understandings of the digital image within the early development of computer vision are indebted to cultural logics as well as technological advancements and limitations. This is to say that the different organizations of the image found within this period, including the classification of discrete yet whole images into binary categories, and a mobile frame of sightless seeing produced by semiautonomous sensing devices, are understandable with the general common sense of a historical moment. Computing as such may, as David Golumbia argues, have its own set of beliefs that "[tend] to be aligned with relatively authority-seeking, hierarchical, and often politically conservative forces," but the specific practices and techniques can also be the product of more historically specific understandings.[42] An instructive example of historicizing computational thinking is Tara McPherson's account of the emerging modularity of mid-1960s computing systems as coinciding with another logic: the lenticular lens created by midcentury three-dimensional postcards that conjoins two images but simultaneously isolates them as the lens device prevents seeing both images together.[43] McPherson sees the forms of the separate but equal lenticular modularity embedded within the surviving computing paradigms that were originally developed midcentury as continuing to shape our thinking and lives in the present, even as they have been revised and updated. "In the post–civil rights United States," she argues, "the lenticular is a way of organizing the world. It structures representations but also epistemologies. It also serves to secure our understandings of race in very narrow registers, fixating on sameness or difference while forestalling connection and interrelation."[44] McPherson's historicization of computational modularity interprets these techniques in relation to a broader cultural logic that touched many aspects of social life. The interdependencies that McPherson identifies between midcentury computing and racial logics shows both the widespread dispersion of such cultural logics and the ways in which they are reproduced by technology. In the case of computer vision, dominant Cold War paradigms and military reconnaissance-driven understandings made certain methods, such as binary classification and the "oppositional intelligence" of cybernetics, sensible as they were being developed, and in turn, these new methods would eventually come to modify perception and reframe the sensible.

Computer vision is thus always more than just technologies of seeing—that is, the algorithms and methods used to automate vision. It involves an approach toward understanding digital representations of reality that is primarily rooted in the analysis of these encoded visual data, but it also draws on other forms of knowledge of the world. Computer vision is also discursive. It is located in code, in systems—technical as well as social—built around specific algorithms, and in the rhetoric articulating this mode of understanding the world as part of visual culture. To understand computer vision is to take on its sense of the world as a particular construct, a particular metaperspective toward reality, one that is shaped by its history. Louise Amoore's concept of "cloud ethics" can help us understand the world-shaping activity of technologies like computer vision by examining these algorithms and methods as ethicopolitical entities: "There is a need for a certain kind of ethical practice in relation to algorithms, one that does not merely locate the permissions and prohibitions of their use. This different kind of ethical practice begins from the algorithms as always already an ethicopolitical entity by virtue of being immanently formed through the relational attributes of selves and others."[45] Amoore's prompting critics to ask how these technologies embody ethics as propensities and properties expands the scope of our analysis from inputs and outputs to the relations formed by their various constructions, assemblages, and embeddings. Algorithms, as we see in the case of computer vision, cannot be conceptualized as having some innate contemporaneity. If algorithms can be said to have ethicopolitical relations, these would be instantiated at their genesis and updated through their subsequent modifications and use.

Joining the critical apparatus of cultural studies with a type of analysis suited to complex technical systems, one attuned to their special limitations and affordances, can illuminate the discursive and technical functioning of computer vision. Because computer vision is always already a complex system, we can approach each stage of constructing the fantasy of sightless vision, a seeing without people, as a unit within a larger assemblage of technology and systems. This enables us to temporarily focus on the functioning of the part without ignoring how it is connected to the whole. Ian Bogost's conception of "unit operations" can bring us closer to articulating the function of both the flowchart of computer vision and the individual blocks within the larger diagram. "In system analysis," Bogost writes, "an operation is a basic process that takes one or more inputs and performs a transformation on it. An operation is the means by which something executes some purposeful action."[46] Turning to unit operations allows us to bracket some of the intricacies of algorithmic transformations while preserving our overall analysis of the complex system. Understanding computer vision as at least partially constructed through the stringing together of units that perform numerical transformations on visual and other data sources enables degrees of scale that can be abstracted to address, for example, the multiple categories of feature extraction that may be composed of elements as concrete as individual pixel groups or as abstract as symbolic representations of probability values. Bogost's unit operations are especially well suited to extensible classes of problems within digital applications, but computational units, especially within the machine learning context, are not always discrete as recursion and self-modifying structures complicate the drawing of boundaries. It is also the case that by depending on the function of some of the units, other units might not be necessary or may even introduce a greater probability for misclassification. Bogost's unit operations are part of his "alien phenomenology," an approach to understanding objects through a flat ontology. In contrast, Yuk Hui's ontology places greater importance on relationality, especially as digital objects move between the individual units that make up the computational milieu.

We can ground our developing framework for interpreting computer vision by examining an abstract application from the 1960s designed to match visual objects with those existing within a collection of already known digital images. This general application, typical both then and now, is composed of several distinct algorithms, procedures, and data sets. The sample or toy procedures—or, to use the more technical term, workflow—used by W. S. Holmes of Cornell Aeronautical Laboratory show the influence of the early coarticulation of machine learning alongside computer vision for the analysis of digitized images.[47] Holmes's workflow begins with the acquisition of data from *sensing* instruments that sample some form of visual stimuli. These data are then moved through a *preprocessing* procedure that might select, quantize, or normalize numerical representations of sensed data. Preprocessing, as Holmes writes, "reduces the dimensionality and grey levels or both prior to the derivation of properties. It may also remove or reduce variations of size and orientation."[48] These values are then supplied as input to a *property derivation* element that assigns meaning to a collection of values by grouping them together into regions or higher-level abstractions of image features. The resulting values are then used as input to a *classification* procedure that learns how to discriminate among the supplied property or feature values. Holmes has as a final stage of his workflow a *training* procedure that introduces an iterative feedback back into the *classification* process by transforming data from correctly classified objects back into the *property derivation* procedure and updating the criteria used for *classification*. It is the presence of the training procedure in particular that signals the presence of machine learning in this workflow. The criteria for making decisions are not explicitly specified;

rather, they are "learned" from provided samples (e.g., images of airplanes on an airfield and images of empty fields) during the training procedure. The operator implicitly imparts the criteria to be used through the selection of samples. This workflow model was originally produced in the mid-1960s to locate image regions representing component objects of interest through the automatic analysis of aerial photographic images. Through this diagram, Holmes charts some of the major problems and some solutions for computer vision, in the process demonstrating the constraints and limitations of these approaches. His workflow is organized into specialized and discrete units that perform a transformation of input data into suitable output formats for the next step. The major design paradigm of such a workflow is modularity. Each of the algorithms or functions contained within these boxes may substitute for the others. What Holmes refers to as "property derivation," for example, can take many different forms. Within the pixel-based ontology of the digital image, it might involve the thresholding and extraction of pixel values from a particular region in the form of a bounding box, as mentioned above. It might involve lower-level features, like overall brightness of the image, or higher-level features that encode elements of the represented scene in symbolic form. The classification stage, to take one of the most common sites of workflow substitution, especially in the form of experimentation with alternative algorithms, might involve swapping a simple, well-known, and now traditional machine learning classifier like k-nearest neighbors (k-NN) for a more complex learning algorithm making use of neural networks. These are likely to produce slightly different results; one might prove more sensitive to values extracted through the property derivation stage, or one might be easier to interpret. Typically such substitutions or changes can be incorporated without altering other units within the larger workflow. The sensing, preprocessing, and property derivation units, for example, most likely will remain static during the evaluation of alternative classification algorithms to reduce the influence of these on classification results.

By coupling Holmes's abstract workflow with a more specific application for computer vision, we can better understand the stakes of alterations within each unit and the need for understanding and critique to be directed toward the entire system and not just at the level of the singular unit operation. The following imaginary application of computer vision appears in the introduction to Richard O. Duda and Peter E. Hart's 1973 textbook on pattern classification. Duda and Hart's example provides a slightly more concrete, if still somewhat fanciful at the time, use of computer vision that can be easily mapped onto the basic operations included in Holmes's block diagram:

> Suppose that a lumber mill producing assorted hardwoods wants to automate the process of sorting finished lumber according to species of tree. As a pilot project, it is decided to try first to distinguish birch lumber from ash lumber using optical sensing . . . The camera takes a picture of the lumber and passes the picture on to a *feature extractor*, whose purpose is to reduce the data by measuring certain "features" or "properties" that distinguish pictures of birch lumber from pictures of ash lumber. These features (or, more precisely, the values of these features) are then passed to a *classifier* that evaluates the evidence presented and makes a final decision about the lumber type.[49]

Duda and Hart, in agreement with the organization of Holmes's block diagram, recognized that some method was needed to render the image data contained within a set of acquired digital images comparable to each other. Imagine the difference between a cropped close-up portrait and one taken at some distance a mere moment later. Both contain almost exactly the same visual representation, the same depicted figure, but cannot be simply matched by comparing the images using all available pixel data. Creating or extracting comparable data is needed for pattern matching. Duda and Hart proposed handling this need with a "feature extractor," a unit that corresponds to Holmes's notion of property derivation. Holmes recognizes the importance of preprocessing, which in this case shows that some modification of the "raw" information in the acquired image is needed in order to extract features or properties from the image. Turning from comparing raw pixel values to standardized properties or features may alter much of the nature of the visual representation; it certainly introduces ontological changes in the status of the compared objects. Likely no longer functioning as symbolic representations of reflected light intensities, these new features may introduce alternative methods of representing the same visual object. These features introduce additional layers of abstraction into the model, and there are no guarantees that the features correspond to what was assumed at the lower levels. As image data move through these workflows, which can be read through Yuk Hui's conceptualization of digital milieu, they are altered and marked by the stamp of a particular operation, yet they remain connected by the history of prior transformations. This understanding of data provenance underwrites both the ontology of digital objects offered by Hui and computer vision's ability to say that while a whole series of transformations can take place on a digital image, the object produced at the end can still be used to make statements about the visual representations of the real world found in the initial object.

The line of separation between computational units in Duda and Hart's abstract model is, as they remind their readers, theoretically arbitrary.[50] Reading particular implementations of computer vision workflows against an abstract model and presupposing the existence of discrete unit operations within this implementation will lead to a misunderstanding of the system. Abstract computer vision models can be used, however, to get closer to the localized problem at hand and to focus our attention on the implemented units. With an understanding of what is present, we can inquire as to the operation of these particular units and seek to understand why these rather than other possible units were selected. We also need to inquire into what precedes and informs the technical units. In Duda and Hart's explication, feature extractors operate in a feedback relation to human observers who use them to propose possible alternative features that are based on different modes of representing these visual objects. The classification algorithms may, in the case of more complex algorithms, also have the ability to directly modify the feature extractors, or they may function in a similar feedback relation to human operators who implement these changes themselves. This is to say that the observational environment should be understood to precede the appearance of the seeing machine that functions, at least at the level of discourse within computer vision, without the assistance of human perception. This is just one way in which the "objective" position of computer vision might be said to be enabled through the encoding of prior subjective decisions made by perceptual humans.

Duda and Hart's model is aware of this liability. They construct the computer vision application as a skeptical reader of its environment. At the same time, their approach demonstrates the degree to which computer vision techniques that deploy pattern recognition are utterly dependent on training in the form of human knowledge of what constitutes valid features and patterns. In the following lumber mill example, they explain how they test the validity, as measured by their assembled workflow of feature extractors and classifiers, of statements made by perceiving subjects within the system that they wish to model:

> Suppose someone at the lumber mill tells us that birch is often lighter colored than ash. Then brightness becomes an obvious feature, and we might attempt to classify the lumber merely by seeing whether or not the average brightness x exceeds some critical value x_0. . . . No matter how we choose x_0, we can not reliably separate birch from ash by brightness alone . . . In our search for other features, we might try to capitalize on the observation that ash typically has a more prominent grain pattern than birch. This feature is much more difficult to measure than average brightness, but it is reasonable to assume that we can obtain a measure of grain prominence from the magnitude and frequency of occurrence of light-to-dark transitions in the picture. Now we have two features for classifying lumber, the brightness x_1 and the grain prominence x_2.[51]

As their example demonstrates, computer vision requires at the very least some minimal human direction for decision-making, but it cannot assume that the provided criteria are either measurable within the technical limitations of the included feature extraction unit or that these criteria are the statistically correct measurements for determining separability among the provided classes of objects. This places computer vision in a double bind. On the one hand, it operates according to what we might call a skeptical evaluation of existing criteria—criteria that can be updated, accepted, or negated according to their own classification accuracy. On the other hand, its design paradigm is shaped according to the limitations of provided knowledge ("that birch is often lighter colored than ash") and prior decisions made through human observation.

Computer vision, even as articulated by its founding researchers, is always a composite technological and social system composed of multiple technical units, existing knowledge of the world acquired by specialists, and a series of decision points. Some of these decision points are explicitly articulated, encoded, and decisively revalued by humans; others are automatically determined, derived from fitting trained data to a model, an operation that might be said to be quasi-quantitative at best, in that the decision logic is the result of encoding a series of perhaps conflicting prior human decisions.

Computer Vision as Visual Culture

The advent of computer vision produced an alteration to the concepts of the image, vision, and perception and inaugurated a new scopic regime. This regime initially emerged from the norms and needs of the mid-twentieth century as an update to visual military intelligence. The familiar scene and apparatus of photography and film had been disrupted by these changes. Created to mimic the human perceptual system, these technologies distrust perception, and they have little use for eyes. Without a self, computer vision is an other observing others. Within this new regime, viewable images and image data have multiple audiences; some of the image data are created for other algorithmic processes and may not even be viewable by humans. Even as they are conceptually understood to be representations of something in the world, within the discourse and methods of computer vision, these images are primarily conceived of as containers or carriers of information. Sight is thus not prioritized in computer vision. It is rather the possible information that can be extracted from sensed image data that is the object of computer vision. They exist outside of a need for a viewing subject; they do not require screens. These techniques operate on images without the need to display images, for the meaning of an image does not reside within the image itself but within transformed image data and models produced through the combination of multiple images. All images within computer vision are subject to the existence of other images.

The developers of several of the major computer vision and machine learning algorithms were trained as military analysts of aerial film photography. These photointerpreters were able to adapt their experiences manually matching patterns found in photographs to data and eventually to digital images. The methods they developed were designed to operate on two-dimensional, top-down aerial surveillance and reconnaissance photographs. These images, because of both demand and expertise, became the primary object of analysis in early pattern matching and computer vision. The "perspective" of early computer vision might be said to be constructed by two different aspects, which is to say two views. The first is directly connected to the scene of image acquisition and aerial imaging. This perspective persists in the present with the immense popularity of consumer-grade aerial drone photography. The second perspective takes place in a different space: a mathematical space that for the purposes of this critique we should understand as virtual. This virtual space is an alternate rendering, a transformation. Taken together, these two aspects produce a view of the world that incorporates all of the imperial power to dominate the landscape found in depictions of the bird's-eye view and later in militarized aerial photography. The perspective of computer vision, however, intensifies the deployment of power over visual space with what might be best described as a will to fit any sensed scene into a newly imagined space. This perspective, originally imagined as operating with the previously invoked top-down view, can reorder sensed data into innumerable representations, altering geometries and coordinate systems in a drive to make what is visible into visual information, to render all data sensible. It was this drive that led the creators of computer vision to propose it as a superior alternative to human vision.

Computer vision is very much a part of visual culture in the present. There is no way to discuss the circulation and manipulation of images in the twenty-first century without engaging in some aspect of computer vision. What Lev Manovich describes as a shift from "lens-based to *computational photography*" encompasses just part of the epochal shift produced by computer vision.[52] Regardless of our understanding or wish to engage with computer vision, our interactions with visual culture, especially when we view, modify, search, and share digitally mediated images, are shaped by these technologies. We are surrounded by imaging devices. With the attendant circulation of digital images and video, our contemporary moment is as image saturated as any vision of postmodernity offered in the early 1990s. The processes and procedures of computer vision can include such familiar applications as optical character recognition and the application of a filter to an image with Instagram. Yet our interactions with computer vision also function outside of our perception of visual images. The surveillance state network making use of computer vision reaches widely and includes privately operated closed-circuit television and traffic light cameras, as well as personal smartphone cameras and official customs kiosks. Our sense of space—from how we now experience flight to our sense of how our automobiles move through the world—has been altered by computer vision, and with this reconfiguration of space, the background by which perception takes place has also been altered. Understanding how the computing past informs the visual culture of the present is essential to comprehending the historicity and function of algorithms that are themselves using knowledge from the past to reframe the present.

« 2 »

Inventing Machine Learning with the Perceptron

The histories of machine learning, computer vision, and Cold War military technology and its application are so intricately knotted together that any attempt to pull these developments and events apart and treat them as individual objects of analysis will result in an incomplete account. While some of the major algorithms and ideas supporting machine learning could be called data agnostic in that they can be used on many different kinds of data derived from all sorts of objects, the early algorithms and the field itself were designed in combination with the explicit goal of the classification of photographic images and the objects contained within these images. This goal was essentially military in nature and was at the origins of what we now recognize as surveillance culture. Machine learning began in a research laboratory with privileged access to an archive of a large number of photographic images, including aerial photographs of landscape features and targets of interest to the U.S. military. These photographic images became natural objects for researchers working on these algorithms, and their features shaped the scope of the problems for which computer vision would become the solution. The dream of machine perception and an automated mode of analysis that could quickly and accurately separate figure from ground, classify and count certain kinds of objects, and determine the military value of a particular site drove the funding priorities and set the targets that motivated computer scientists to develop, in the 1950s and 1960s, systems and software that laid the groundwork for a more generalized computer vision.

While "artificial intelligence" and "machine learning" are frequently taken as interchangeable terms and have overlaps in the communities that describe their work using these words, they have been articulated differently at different times since their emergence. The methods and especially the discourse of artificial intelligence have gone in and out of fashion, depending on the public perception of these techniques' efficacy. The term "artificial intelligence" was invented for the purposes of a ten-week summer conference, the Dartmouth Summer Research Project on Artificial Intelligence, held at Dartmouth College in 1956. John McCarthy, a newly hired assistant professor of mathematics, hosted the conference in Dartmouth Hall, an imposing Georgian building at the center of the small campus. McCarthy invited leading researchers including Oliver Selfridge, Claude Shannon, Ray Solomonoff, and Marvin Minsky to discuss, among other topics, machine creativity, neuron nets, and machine-based self-improvement. Although the Dartmouth workshop responsible for coining the term "artificial intelligence" was directed, as its commemoration plaque in Dartmouth Hall notes, "to proceed on the basis of the conjecture that every aspect of learning or any other feature of intelligence can in principle be so precisely described that a machine can be made to simulate it," a far more basic goal motivated the researchers who would develop the initial machine learning approaches.[1]

We are now accustomed to using the phrase "machine learning" to cover a large number of statistical approaches to the problem of data classification. Although the phrase has entered into the lexicon of the general public as almost a synonym with artificial intelligence, "machine learning" has a genealogy distinct from "artificial intelligence." The statistical approaches that we now classify under the rubric of machine learning include linear discrimination functions that were invented in the 1930s and statistical rules for nearest neighbor detection first articulated in the early 1950s.[2] Neither of these methods requires large amounts of data or even digital computers. Training data function as the criteria by which one might determine that a particular algorithm or approach utilizes machine learning. The public's response to artificial intelligence appears like the waves of a sine curve, moving from excitement and hype to disappointment and rejection, then back to excitement. At several moments in the past, when it had become no longer fashionable or politically viable to be associated with artificial intelligence, many researchers redefined their projects. Computer vision is one such project. Machine learning and artificial intelligence are two distinct discursive constructions that depend as much on the vicissitudes of cultural dispositions, the changing uses and different connotations of these terms within government agencies, corporate marketing, and the media as they do on the structured logics internal to the algorithms.

Yet at the same time machine learning has a specifically material history. The researchers exploring ways to automate perception and recognition frequently built special-purpose physical computing devices to perform these tasks. The algorithms that they developed were initially embedded in hardware, so we might say that software implementations of machine learning algorithms are virtual emulations of mechanical devices, themselves frequently modeled on biological systems. While many machine learning methods were inspired by animal models, in particular the visual systems of frogs and cats, their manifestation as material objects formed the possibilities and linguistic legacies found even within contemporary implementations.[3] References to these analog computers—the machines of machine learning were imagined and materialized as physical devices with highly specialized and custom hardware—are found throughout the conceptual language of machine learning. The operation of these machines, the parents of today's convolutional neural networks and deep learning algorithms, were shaped by and dependent on midcentury binary thinking and behaviorist biological theories. The activities for which these early machine learning devices were designed are also imprinted on the legacy of these methods, for the development of these machines and their algorithms were directed toward one very specific task: the automation of photointerpretation for surveillance and military reconnaissance.

The Cornell Aeronautical Laboratory (CAL) in Buffalo, New York, was the birthplace of machine learning's most famous early neural network algorithm: the Perceptron. CAL had once been known as the Curtiss-Wright Aircraft Corporation, a private research company dedicated to aircraft research and highly active during World War II.[4] After the conclusion of the war, in 1945, the Curtiss-Wright Aircraft Corporation gave their research facility to Cornell University, which is located more than 150 miles away, in Ithaca, New York, "in the interest of aviation and the welfare of the nation," with the corporation calling the laboratory "an instrument of service to the aircraft industry . . . to education . . . [and] to the public at large."[5] The university formed CAL in 1946 to continue aviation research at this Buffalo location. Much like other university-affiliated research laboratories conducting military research during the Cold War and during the Vietnam War, CAL was eventually forced to sever its relationship with Cornell University in response to growing student protests over applied research that was being deployed by the military. In 1972, Cornell turned CAL back into a private company, calling the new organization the Calspan Corporation. During the 1950s and 1960s, CAL provided an institutional home—and, more importantly, access to U.S. Department of Defense funding—for a group of researchers interested in advanced statistics, perception, and the possibilities of automated photograph analysis.[6]

The inventor of the Perceptron algorithm, Frank Rosenblatt, was a psychologist trained at Cornell University. Rosenblatt earned his AB degree in 1950 with a major in social psychology and a PhD from this same institution in psychology in 1956. He studied anxiety and personality patterns in relation to parenting practices, and he also conducted research on schizophrenia with the support of the U.S. public health service. Rosenblatt started working at CAL as an engineer before concluding his graduate studies at Cornell, and would continue to do so, as head of CAL's Cognitive Systems Laboratory, before returning to the Ithaca campus to take up a position as a lecturer in psychology and director of the cognitive systems research program. In 1965, he was promoted to associate professor in Cornell's division of biological sciences.[7] Rosenblatt combined insights across numerous fields and made contributions to multiple disciplines. This was evident from his earliest work. His dissertation was titled "The k-Coefficient: Design and Trial Application of a New Technique for Multivariate Analysis."[8] While describing new statistical methods for the analysis of data, his dissertation fit into the field of psychological research. His dissertation describes a computational method for addressing multiple correlations across a large number of variables, an activity that was becoming increasingly common in the survey-based psychological studies of the mid-twentieth century. Rosenblatt was especially interested in applying his newly developed technique to behavioral studies in personality research, social psychology, and psychopathology.

The Perceptron, Rosenblatt's best-known contribution to the psychological and brain sciences, might best be thought of as occupying the transitional space within the field of psychology that was emerging between behaviorist and neuroscientific accounts of the mind. His dissertation research demonstrates the degree to which this mid-twentieth-century moment was in theoretical and methodological flux. While his dissertation was a behavioral study grounded in the dominant theories of the time, he was already interested in computer-based brain models. In order to test his hypothesis for the experimental statistical method that would make up his dissertation project, Rosenblatt needed to collect large numbers of survey responses with a large number of questions. Potential participants were told that the study would "determine the relationship of family background to students' future plans and problems."[9] The survey asked participants to report on scenarios that involved the adult men and women who took the most responsibility for the participant during their first twelve years. Many of the questions were given in paired form using the structure of "always" and "never" linked to specific behaviors, requiring the participant to choose between two radically opposed options. The following are representative questions from sections addressing behaviors of the subject's caregivers:

> 8-17. My mother almost never punished me for anything that I did (not counting ordinary criticism); she probably punished me less than once a year.
>
> My mother was always punishing me for something, practically every day, during my first twelve years.
>
> 11-28. If I did something he didn't like, my father would keep reminding me of it long after it was over rather than letting it drop; he was always holding things I had done over me.
>
> He almost never brought up things I had done wrong, and might be unable to recall them even when he was reminded afterwards.[10]

Rosenblatt's survey questions probed the various ways parents may have corrected their children's behavior. While the questions provided room for subjects with a nonnuclear family, the survey was designed around mid-twentieth-century gender norms and rather stark understandings of what constituted normative parental behavior. He appeared especially interested in tensions between parents and children. The basic assumptions, the understanding that the future success of individual subjects was a result of their home environment, were built on the norms of behavioral psychology.

Rosenblatt's solution to the problem of addressing these many variables that registered a range of responses across his large subject pool involved the creation of both a statistical method, what he called the k-coefficient, and the design and construction of a special-purpose digital computer, the EPAC, or electronic profile analyzing computer. In addition to the development of this computer, Rosenblatt used a general-purpose computing machine, an IBM card-programmed computer owned by the Cornell Computing Center at Cornell University, and the Datatron computer at CAL. Rosenblatt's custom EPAC design was partially a simplified adaptation of the design used for the ENIAC, or electronic numerical integrator and computer, that had been built earlier for the University of Pennsylvania. The EPAC was not a general-purpose digital computer. It did not have any data storage devices. It had no software. There were no displays. Its input capability was quite limited. A drum processed not the punch cards that were typically used as input during this period but paper forms containing responses to his surveys. The EPAC simply read two columns of marks produced with an electrographic pencil and stored in an accumulator the squared differences of the two columns.[11] Rosenblatt called his device an "idiot brain," which gestured toward his desire to develop a machine that would more closely simulate aspects of a biological brain.[12] The EPAC illustrates Rosenblatt's ingenuity in solving complex tasks and his willingness to design and create custom computing hardware. Creating specialized computational machines for a single purpose was a much more common practice at this time. Specialized hardware enabled the use of new experimental methods and gave researchers privileged access to computing hardware. Access to computers was crucial because the general-purpose computers used by Rosenblatt for his dissertation work were shared with many other researchers, both at CAL and at Cornell University, and the construction of the EPAC gave Rosenblatt dedicated access to a device capable of processing some of his data locally and without waiting. The notion of developing specialized computing devices remained with Rosenblatt as he launched his postdoctoral research projects at CAL and began work on what would become the Perceptron algorithm and its specialized hardware implementation as the Mark I Perceptron.

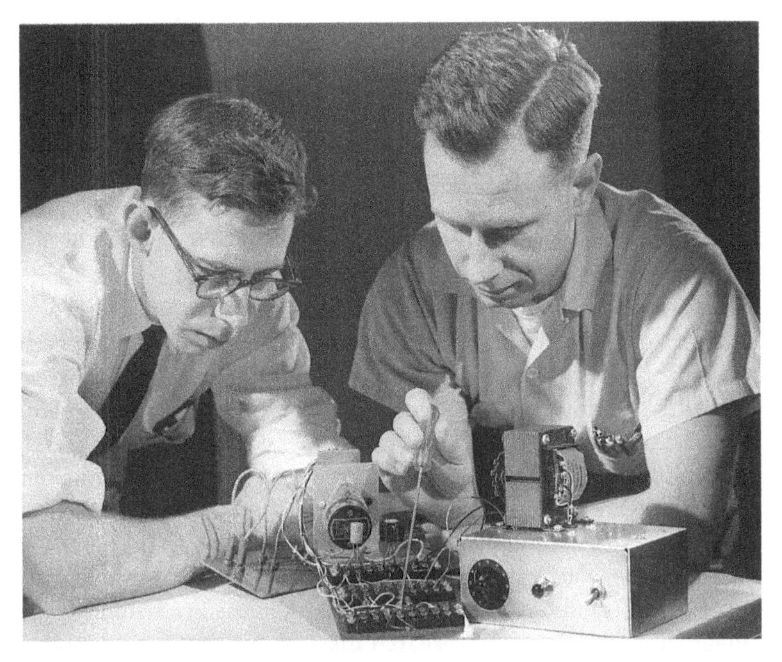

Figure 3. Frank Rosenblatt (left) and Charles W. Wightman working on a prototype association unit for the first Perceptron, December 1958. RMC2005_0771, Cornell University Faculty Biographical Files, #47-10-3394. Division of Rare and Manuscript Collections, Cornell University Library.

In January 1957, Rosenblatt published a short technical report titled "The Perceptron: A Perceiving and Recognizing Automaton (Project PARA)." It is apparent from this initial report and the accompanying funding proposal that the mode of perception of concern to Rosenblatt would be primarily visual. Rosenblatt's report introduced the concept of the Perceptron as a generic class of learning machines and provided some initial designs of the algorithm as well as its proposed material implementation as a specialized machine, along with a larger, multidevice system, which he called a photoperceptron, that would be used for what he considered to be the major target applications for his invention. Rosenblatt's report builds on his dissertation research in several fundamental ways. One primary lesson that he takes from his research in behavioral psychology is that the presence of humans in the recording and processing of data is an unnecessary intervention in experimental research. Rosenblatt's invention was humanlike but closer than humans could ever be to observed phenomena. "Interest has centered on," he writes in the passive voice of the scientist, "the idea of a machine which would be capable of conceptualizing inputs impinging directly from the physical environment of light, sound, temperature, etc.—the 'phenomenal world' with which we are all familiar—rather than requiring the intervention of a human agent to digest and code the necessary information."[13] The report builds on his prior development of statistical methods to handle many variables, the k-coefficient, and his ongoing professional need to categorize or discriminate data. Rosenblatt's proposed algorithm drew on recent highly influential models based on physiological accounts of brain function—most importantly, the transmission of excitations as nerve impulses, which is to say electrical signals, from one neuron to another in a dense network that, when taken together, were understood to provide a coded or symbolic representation of the stimulus response.

The name "Perceptron" etymologically combines the two concepts that Rosenblatt hoped to synthesize in his invention. It was imagined as an electrical research instrument that would enable experiments in understanding perception of the world. The "Perceptron" name borrows from physics and the cyclotron (1935) the sense of a specialized experimental device that enables researchers to examine otherwise difficult-to-study phenomena under known and reproducible conditions.[14] Excitation and inhibition are the main biological processes drawn from neurophysiology for which Rosenblatt designs a digital analog in the Perceptron. The three layers of the three distinct systems in his proposal—the sensory (S), association (A), and response (R) systems—are joined together by inhibitory and excitatory connections. These are referred to as layers because although there are many interconnections passing signals between systems, each system is distinct and typically performing some transformation (that is, summation) on the output of the previous layer. Groupings of inhibited and excited A and R units are bound together in what he calls association sets through the activation of specific responses. These association sets form the S layer in the system and function as a memory store of prior responses. Both positive and negative examples are required in order for the S unit association sets to be able to discriminate between categories. The number of appropriate examples is linked to the structure of the entire system. A state of equilibrium of stored associations, a form of data saturation, can occur in which there will be a "loss of previously acquired associations. . . . [and] as this equilibrium condition is approached, it will become harder and harder to 'teach' the system new associations without knocking out old ones, and if we try to re-establish the old ones, we will find that we have lost others in the process."[15] The preservation of prior data was understood by Rosenblatt to be a form of memory storage on the model of the human brain. But what was stored? The Perceptron, as designed by Rosenblatt, does not store the data presented to the machine as such, but rather the responses to training examples in the form of activated association sets. What this means is that the Perceptron "learns" through the collection of these forced responses to training examples. These collections are memory traces and could be reconstructed "much as we might develop a photographic negative."[16] In his initial Perceptron model, patterns of A and R unit activations would only work with two categories of data. Extending his earlier work on statistical separability, this design enabled Rosenblatt to construct a system that could discriminate between two already existing categories.[17]

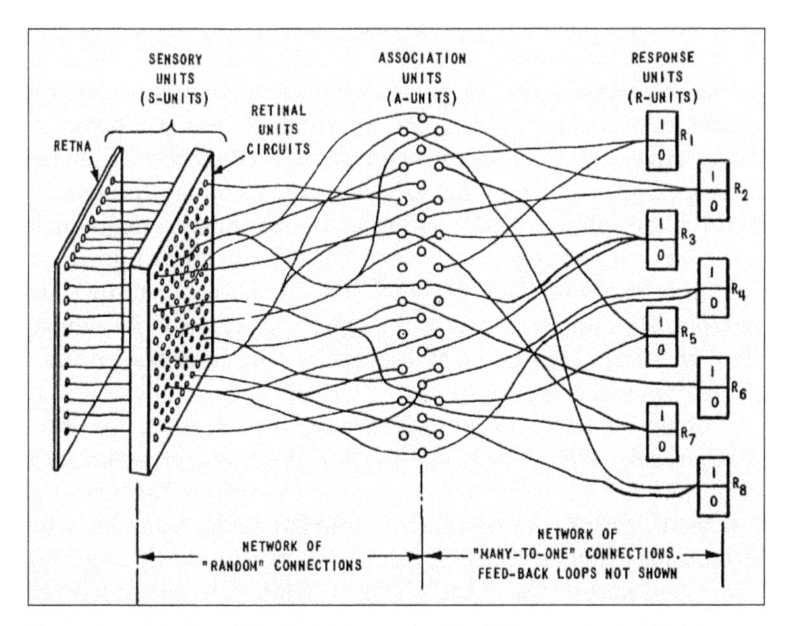

Figure 4. Organization of the Mark I Perceptron (*Mark I Perceptron Operators' Manual,* 1960).

Rosenblatt, like other early proponents of neural nets, was influenced by Warren S. McCulloch and Walter S. Pitts's model of the nervous system. McCulloch and Pitts introduced their model in a 1943 paper titled "A Logical Calculus of Ideas Immanent in Nervous Activity."[18] Their model of what became known as the McCulloch-Pitts neuron was a highly simplified mathematical account of brain activity. The model was organized around binary or two-state values of the neuron; it was either activated or not. Their paper argues that "all psychic events have an intentional, or 'semiotic,' character" that enabled, in theory, an understanding of mental representation of objects by tracing the history of activity in terms of state changes of the neurons within the net: "Specification of the nervous net provides the law of necessary connection whereby one can compute from the description of any state that of the succeeding state, but the inclusion of disjunctive relations prevents complete determination of the one before."[19] This model was highly influential and inspired a whole host of new ways of thinking about computation and simulating brain activity. It provided scientists with a way to model complex activities with a series of simple units. It was also bold. "With determination of the net," McCulloch and Pitts write, "the unknowable object of knowledge, the 'thing in itself,' ceases to be unknowable."[20] Rosenblatt, a trained psychologist, recognized some of the limitations of this model, in particular its atomistic approach, from a psychological perspective, but nonetheless, he found the McCulloch-Pitts model a compelling resource for his experiments in simulating a human brain.[21] The concept of simulation is foundational to neural networks and computing as such. Alan Turing's abstract or universal computer is a simplified computer that computes anything computable. Machine learning, from its birth, is precisely the simulation, on a digital computer, of a specialized machine designed as a simulation of a brain.

Donald O. Hebb's account of learning in perception, *The Organization of Behavior: A Neuropsychological Theory* (1949), was also an important influence on the design of the Perceptron. Hebb's model provided a framework by which Rosenblatt could frame his algorithm as learning, if learning was understood as slowly acquiring perceptual knowledge of the world. In relation to the process of perceptual learning, Hebb offered two related propositions. First, at the neural level, learning involves the repeated activation of patterns of particular cells. Second, people learn to perceive figures not as wholes but through an additive process of reconstructing previously recognized parts of figures. Hebb's account of perception, which contested some aspects of the gestalt theory of perception, was that "quite simple diagrams are not perceived directly as *distinctive* wholes—that, though the stimulus has a unitary action in the figure–ground relationship, the perception of identity depends on a series of excitations from the parts of the stimulating diagram."[22] Unlike some theories of regional specialization, in which certain groups of neurons would be activated by certain visual stimuli in their entirety, Hebb's notion was that information about the parts or components of stimuli would be encoded by specific neurons appearing throughout the network of neurons. There was also a degree of randomness in Hebb's model that would be incorporated in the design of the Perceptron. The Hebbian model of learning was especially attractive to researchers working with simplified models of neural nets and with far fewer artificial neurons. It also enabled Rosenblatt and others to propose general-purpose pattern-matching systems that could learn to differentiate patterns from data representing many different types of objects—even beyond visual objects. This was in part because the Hebb model was theorized as a form of nonspecialized learning.[23]

The machine learning algorithms that implemented what we might think of as a universal form of learning as found in Hebb's model were well suited to the dominant ontology of computer vision at the time. This ontology assumed the presentation of a series of images as a complete pattern of elements or pixels. Pattern detection within this paradigm involved no specialized detection units, only repeated samples of the entire image. The "space" of the image would later be fractured, and specialized feature detection components would be added to the machine learning repertoire. This shift from universal to specialized learning was also taking place within psychology and physiology. McCulloch and Pitts, joined with Jerome Lettvin and Humberto Maturana, would in 1959 publish a highly influential analysis of the frog visual apparatus titled "What the Frog's Eye Tells the Frog's Brain." This paper provides an analysis of a biological model of nerve activity connected to vision that supplements the stripped-down McCulloch-Pitts neuron. Using data collected by recording signals from various fibers in the optic nerve of a frog, the authors propose the existence of four detectors for different categories of stimuli: contrast, convexity, moving edge, and dimming.[24] Taken together, these higher-level detectors would provide the frog with the ability to sense different kinds of visual objects corresponding to important components or features rather than directly sensed and recalled patterns of sensory activity.[25] This model produced a substantial revision of prior neurophysiological accounts of perception. In this model, there was no longer a one-to-one mapping between what is sensed and what is transmitted to the brain. As the authors write, "The eye speaks to the brain in a language already highly organized and interpreted, instead of transmitting some more or less accurate copy of the distribution of light on the receptors."[26] In time, this model of vision based on higher-level features rather than grids of pixel-intensity values would come to be favored by computer vision researchers as well. This will involve, as outlined in chapter 1, a second major ontological shift in the understanding of the image in the early history of computer vision.

There were other biological and psychological influences on the Perceptron as well. Rosenblatt, as previously mentioned, was inspired as much by his behavior research as he was by emergent neuroscientific accounts and mathematical models of neurons and nerve networks. While Rosenblatt was pulled in a biological direction for the organization of his proposed machine, his 1957 technical report and his later conceptualization of machine learning also draw on his behavioral research. His understanding of learning as a behavior, specifically as operator-reinforced behavior, comes from the psychological theory supporting his earlier project. Rosenblatt imported into his design a reward–punishment reinforcement model from behavioral psychology to characterize the positive and negative association set of responses to binary stimulus inputs stored in the Perceptron's S unit memory. He provides a detailed description of the role of the teacher in reinforcement learning:

> It is possible to teach the system to discriminate two such generalized forms, or "percepts," by presenting for each form a random sample from the set of its possible transformations, while simultaneously "forcing" the system to respond with Output 1 for Form 1, and Output 2 for Form 2. For example, we might require the perceptron to learn the concepts "square" and "circle,"

> and to turn on Signal Light 1 for "square," and Signal Light 2 for "circle." We would then proceed to show the system a large set of squares of different sizes, in different locations, while holding Light No. 1 on, thus "forcing" the response. We would then show a similar set of circles, while holding Light No. 2 on. If we then show the perceptron any square or any circle, we would expect it to turn on the appropriate light, with a high probability of being correct.[27]

In this description, Rosenblatt imaginatively turns himself into the corrective parent of his invention, the Perceptron. He instructs the machine on how to turn percepts into precepts. In one mode of operation, the Perceptron might be thought of as a digital Skinner box, the laboratory instrument used in animal conditioning experiments. Rosenblatt describes this hypothetical mode of interaction with the Perceptron through which it is "taught" to recognize a pattern:

> The perceptron is exposed to some series of stimulus patterns (which might be presented in random positions on the retina) and is "forced" to give the desired response in each case. (This forcing of responses is assumed to be a prerogative of the experimenter. In experiments intended to evaluate trial-and-error learning, with more sophisticated perceptrons, the experimenter does not force the system to respond in the desired fashion, but merely applies positive reinforcement when the response happens to be correct, and negative reinforcement when the response is wrong.) In evaluating the learning which has taken place during this "learning series," the perceptron is assumed to be "frozen" in its current condition, no further value changes being allowed, and the same series of stimuli is presented again in precisely the same fashion, so that the stimuli fall on identical positions on the retina. The probability that the perceptron will show a bias towards the "correct" response (the one which has been previously reinforced during the learning series) in preference to any given alternative response is called P_r, the probability of correct choice of response between two alternatives.[28]

Reinforcement learning is thus producing at the association level the correct response for the presented stimuli. Parent and child, operator and machine, punish and reward, correct and incorrect—these define the Perceptron's operation and world view.

The binary logic embedded in the simplified Perceptron rendered it incapable of addressing ambiguity. This was not exactly the critique made by other computer scientists, who would find faults in the Perceptron, but the algorithm's requirement for binary problems made it unable to address complex data as well as situations irreducible to separation into two classes. We might consider this limitation a digital manifestation of Cold War logic that required categorical divisions of information, people, and nations. Binary and linear classification, of course, predated the Cold War; early examples include Ronald Fisher's classic 1936 discriminating linear function detailed in "The Use of Multiple Measurements in Taxonomic Problems"—but its logics were highly compatible and comprehendible within the organizing cultural frame of this historical moment. The linear separability between two categories, especially "them" and "us," was one of the primary governing logics of Cold War culture in the United States. Computer historian Paul N. Edwards describes how the "closed world" produced by containment, the central and widely used keyword of the Cold War, permeated the culture of this period:

> On one reading the closed world was the repressive, secretive communist society, surrounded by (contained within) the open space of capitalism and democracy. This was the direct intent of the containment metaphor. But on another reading, the closed world was the capitalist world-system, threatened with invasion. It required defenses, a kind of *self*-containment, to maintain its integrity. In the third and largest sense, the global stage as a whole was a closed world, within which the struggle between freedom and slavery, light and darkness, good and evil, was constantly joined in every location—within the American government, its society, and its armed forces as well as abroad.[29]

Simple binary categories, such as differentiating squares from circles, were the test cases for the Perceptron. As the algorithm developed, new applications were proposed that expanded the categories of classification to include the absence or presence of aircraft in revetments as well as other binaries, such as learning to differentiate between men and women in photographs. The Perceptron's design worked well within what we might think of as the conceptual space afforded by the Cold War frame, but when actually applied to problems arising from the military-defined goals during this moment, it was subject to numerous limitations.

The Machines of Machine Learning

The Mark I Perceptron was the materialized implementation of the Perceptron algorithm. It was completed by February 1960 and resided in Buffalo at the CAL. The Mark I, as a mechanical computing device, would imaginatively take the place of what was previously a virtual or simulated "machine" in Rosenblatt's account of his Perceptron algorithm. It was large—larger even than many of its contemporary general-purpose digital computers. It was not operated with traditional input devices or punch cards but directly with switches on the device itself. The Mark I was built to discriminate between classes of objects from supplied visual input. It was a linear classification machine. It "learned" to draw a line, what is called a hyperplane, between high-dimension data in order to separate the researcher-presented data into two preestablished classes or categories. The publication of a mathematical proof that the Perceptron would eventually "converge" on a solution—which is to say it would find the correct hyperplane by which it could separate two classes, if they could be separated—was one of the algorithm's major accomplishments.[30]

The Mark I was the material manifestation of Rosenblatt's long-held dream to produce an alternative to the general-purpose digital computer as what he called a "brain analogue" for perception of the phenomenal world.[31] It was constructed by Frank Rosenblatt with the assistance of Charles W. Wightman, among several others. In their operators' manual, the system's designers describe their invention as "a pattern learning and recognition device. It can learn to classify plane patterns into groups on the basis of certain geometric similarities and differences. . . . Among the properties which it may use in its discriminations and generalizations are position in the retinal field of view, geometric form, occurrence frequency, and size."[32] The machine was equipped with a camera and the S units were connected to photoresistors. There were four hundred photoresistors in the Mark I, mounted in a 20 × 20 array. These photoresistors, coupled with the S units, formed the "retina" of the Perceptron. Because these light sensors were the only sensory input to the Mark I Perceptron, it would be operated as a photoperceptron.

The operators' manual for the Mark I was published in February 1960, not long after Rosenblatt had accepted a research position and faculty appointment at Cornell University. Despite the complexity in designing and building the Mark I, like most computers—both conventional digital computers and specialized hardware—it did not have a long life. Rosenblatt remained involved in research addressing the Mark I Perceptron project after leaving CAL and continued work on simulated Perceptrons for several more years. The first major application project for the Mark I Perceptron was called Project PICS. This research was funded by the Office of Naval Research and focalized research on the device in the direction of pattern recognition of aerial photographic images of "complex, militarily interesting objects."[33] The first report on phase 1 of Project PICS was published in November 1960. The system would also be used in several other experiments by CAL researchers working under Albert E. Murray, who had previously been Rosenblatt's supervisor at CAL, and W. S. Holmes. The complete Mark I Perceptron system was eventually relocated to the Smithsonian National Museum of American History.[34]

There were other physical machines created on similar models as the Mark I Perceptron. In 1959, Albert Maurel Uttley of the National Physical Laboratory in Middlesex, England, developed what he called a "conditional probability computer" as a hydraulic, which is to say mechanical, computer. The memory storage of this "special purpose computer" was in the form of liquid levels in self-contained vessels controlled by the system's hydraulic mechanism.[35] Designed to imitate animal learning, Uttley's device demonstrated a method of binary classification using Bayesian probabilities. He used a biological model of response training that also participated in mid-twentieth-century behaviorism through his conception of memory as requiring recovery and reflex: "If, after a long period of conditioning an animal, the conditioned reflex is extinguished by nonreinforcement of the unconditioned stimulus, the reflex will then recover spontaneously even if no further conditioning takes place."[36] Researchers at the Stanford Research Institute created two physical machines, MINOS and MINOS II, with funding from the U.S. Signal Corps, "as part of a research program in the

Figure 5. The Mark I Perceptron at the Division of Medicine and Science, National Museum of American History, Smithsonian Institution.

recognition of graphical patterns."[37] In Italy, Augusto Gamba designed a mechanical learning machined called PAPA (rendered in English as Automatic Programmer and Analyzer of Probabilities) in 1961 at the University of Genoa.[38] Like the Mark I Perceptron, Gamba's PAPA had an array of photocells that sensed light and used these values (on or off) as patterns of excitation for classification. The PAPA was especially designed for pattern recognition of photographic images. Gamba would later develop a smaller system based on the PAPA that he called the Papistor for binary image classification using optical fibers instead of photocells.[39] These two examples demonstrate the broader interest in learning machines and image recognition.

While others had developed their own learning machines to experiment with these simulated animal learning models, Rosenblatt's continued involvement in the field was now mostly attached to the Tobermory, a follow-up project to the Mark I Perceptron. Like the Mark I Perceptron, the Tobermory was constructed as a physical machine. It was one of a number of designs for analog memory devices based on the Perceptron model.[40] The Tobermory was created at Cornell University and was the subject of George Nagy's 1962 dissertation, titled "Analogue Memory Mechanisms for Neural Nets." Nagy was advised by Rosenblatt and carried out his research in collaboration with others in the cognitive systems research program. The Tobermory integrated new thinking on an analog memory or storage device called the "magnetostrictive readout flux integrator," developed by Charles A. Rosen of the Stanford Research Institute.[41] Rosen's innovation was crucial to the design of the Tobermory and an improvement on the memory device used in the Mark I Perceptron. This was a much larger system than the Mark I Perceptron. It was a four-layer network and had 12,000 weights and two layers of 2,600 A units to the 512 single-layer A units found in the earlier machine. Despite Rosenblatt's expansive list of potential applications, the Perceptron was designed as a photoperceptron; the Tobermory, as a major revision of the concept in mechanical form, was designed to perform pattern recognition on audio data. This task and the biological model inspired the name of this device: "Tobermory, named after Saki's eavesdropping talking cat, is a general purpose pattern recognition machine roughly modelled on biological prototypes."[42] Despite the claim that the Tobermory would be a general-purpose device, Rosenblatt and Nagy would refer to it as an audio Perceptron. Much like the Mark I had only photoresistors as sensory inputs, rendering it a photoperceptron, the Tobermory had only a microphone or tape head as input devices that would sample audio into 1,600-bit time–frequency–amplitude patterns for classification.

Envisioning the Perceptron

While Rosenblatt describes a world of rich possibilities for the recognition of patterns in optical, electrical, or tonal information, it was ultimately the digital representation of optical data that most interested him and his funders. Rosenblatt's early description of the Perceptron demonstrates the degree to which he understood it as a complete system: "We might consider the perceptron as a black box, with a TV camera for input, and an alphabetic printer or a set of signal lights as output."[43] The military applications and the refining of the Perceptron system's capabilities for identifying objects from military surveillance imagery increased after Rosenblatt's departure from CAL, but it was clear from the initial proposals that his machine was designed with these goals in mind. The Project PARA report explicitly states that "[devices] of this sort are expected ultimately to be capable of concept formation, language translation, collation of military intelligence, and the solution of problems through inductive logic." Rosenblatt's research described in "Perceptron Simulation Experiments" was supported by the Information Systems Branch of the Office of Naval Research.[44]

After Project PARA and the implementation of the Perceptron algorithms, the CAL researchers began work on a more focused project that would be able to perform pattern recognition on visual data. In parallel with continuing work on the Mark I Perceptron, CAL researchers developed techniques for some basic photo-interpretation activities. This required the development of hardware to acquire or sample and quantize photographic images and the use of a Perceptron algorithm on CAL's IBM 704 computer. The engineers used a digital input system that was provided with the 704 to read sampled and encoded data provided by an analog-to-digital device that was attached to the output of a commercial facsimile transmitter. This customized configuration provided CAL with a rudimentary black-and-white scanner that could acquire photographic images at a resolution of a hundred lines per inch, then quantize this information to sixteen levels of gray-scale digital values. It took some time to scan an image—ninety seconds for an image 5 × 5 inches in size—and there was no comparable system for printing as they had for scanning, with image output limited to four rather than sixteen levels of gray used internally. Despite these limitations, the engineers were able to scan images and store digitized information (on magnetic media).[45]

In a demonstration recorded by the BBC and later used in the WGBH series *The Machine That Changed the World* (1992), the Perceptron was used to classify images as either male or female. The binary and linear classification system fit this highly simplified understanding of gender presentation; further, the demonstration reinforced the "naturalness" of gender despite potentially confusing features.[46] The operator presents a series of training images and selects toggles for either man or woman along with either right or wrong to provide positive and negative samples for reinforcement training. After presenting several samples, the operator begins testing. While he presents a photograph of a shaggy-haired member of the Beatles (George Harrison), the narrator explains that the Perceptron interprets facial features and the hairline, and that this image was taking longer than normal. After a moment, a

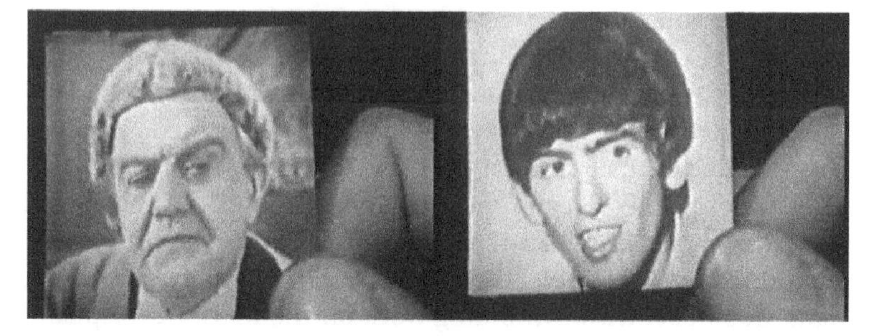

Figure 6. Still frames of George Harrison and Andrew Cruickshank used for the Perceptron experiment. *The Machine That Changed the World,* episode 4, "The Thinking Machine," directed and written by Nancy Linde, WGBH, January 1992.

light behind the sign man is activated. The operator then presents an image of the actor Andrew Cruickshank wearing a wig while the narrator comments on the "heart searching" taking place within the Perceptron before this image is also correctly classified as man. The demonstration uses this advanced piece of computing hardware to restore order to gender representation (Figure 6). George Harrison's appearance cannot deceive the Perceptron, and the classification of him as man functions to show that his presentation is not a threat to a naturalized ordering system that securely categories people as either man or woman.

Interest in gender classification using these new methods was not limited to this video demonstration. Charles A. Rosen used a similar problem—the classification of boys and girls into a binary gender system—initially based on a single measurement or feature, the length of hair, in his explanation of "adaptive" networks in an essay published in *Science* in 1967. Adaptive machines were opposed to fixed machines with preestablished criteria for classification as a "fixed internal organization."[47] These new systems, Rosen writes in a concise description of early machine learning, would alter their organization in response to stimulus:

> A machine composed of adaptive or alterable networks would automatically and progressively change its internal organization as a result of being "trained" by exposure to a succession of correctly classified and labeled patterns. In this case there is no prior knowledge of the relative importance or distribution of the features comprising each pattern. After a training or

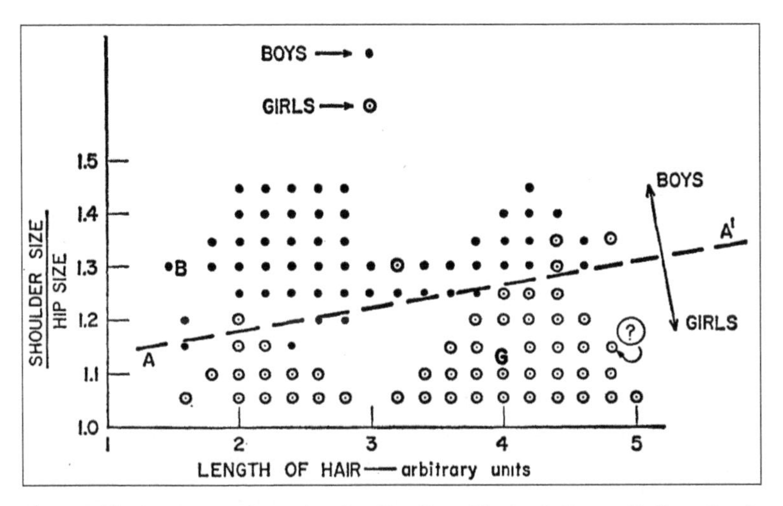

Figure 7. Charles A. Rosen's gender classifier. From Charles A. Rosen, "Pattern Classification by Adaptive Machines," *Science* 156, no. 3771 (1967): 38–44.

> learning process, the final organization would effect, ideally with minimum error, the classification of patterns that are not contained in the training set but are quite similar to them.[48]

The use of hair length as a single feature for classification was especially difficult, Rosen writes, because "a small number of boys today consider long hair fashionable, and some girls prefer short hair." Rosen then adds another feature for each individual—the ratio of shoulder to hip size—and suggests that pitch of voice could also be used to create a "highly idealized representation" of the separating line or hyperplane that would be used by adaptive machines, of which Rosenblatt's Perceptron was the first.[49]

Defining the Photointerpretation Task

Albert E. Murray, a principal research engineer at CAL, published a version of a talk given at the Photogrammetric Engineering Society's annual meeting in March 1961 in Washington, D.C., under the title "Perceptron Applications in Photo Interpretation."[50] In this paper, Murray is careful to distinguish between the general Perceptron algorithm and the image analysis applications that he discusses; like Rosenblatt, Murray understands the Perceptron to have potential application beyond visual pattern recognition, even if these other applications had yet to materialize. He also appears to be concerned, at least within this public presentation, with not representing the Perceptron as too tied to specific military applications. A footnote clarifies that his use of the term "target" in describing the operation of pattern-matching techniques is not intended to indicate only military uses, suggesting some awareness that others might understand photointerpretation and the Perceptron's method itself as military technology: "The term 'target,' as used in this paper, is intended to replace the phrase 'one (kind of) object, feature, topography or terrain' and is to indicate the object, etc. as one for which we wish a recognition machine to give an identifying response or alarm. It is not intended as a military term."[51] Murray is aware that his audience in the Shoreham Hotel might see the pattern matching of a target image as the simulation of and preparation for another kind of target practice. In his initial report to the Office of Naval Research, the funders of CAL's Project PICS, produced a year before his talk at the Photogrammetric Engineering Society meeting, Murray makes clear the intended goal of the Perceptron photointerpretation experiments: "A simple but worthy automation goal would be the ability to scan large quantities of mostly uninteresting photographic material in search of just one kind (or a small number of kinds) of object. Reasonable examples of this mission would be locating ships at sea, or missile or radar sites in the arctic. These are perhaps among the easiest of the worthwhile tasks."[52] Photointerpretation, and in particular military photointerpretation of aerial photographs, was seen as the primary application and testing ground of CAL's photoperceptron.

In his early and influential paper on the computerization of photointerpretation, CAL computer scientist William S. Holmes writes that the task of these human photointerpreters was to "find aircraft, military emplacements, industrial complexes, missile sites, vehicles, storage dumps, etc."[53] This task was to become increasingly complex in the mid- to late 1960s, as the United States ramped up its war in Vietnam and newly developed high-altitude reconnaissance aircraft collected more and more aerial photographs. Holmes's earliest contribution toward the effort to automate photointerpretation was published in a paper titled "Design of a Photo Interpretation Automaton" in 1962. This collaboratively authored project—he published with H. R. Leland and G. E. Richmond, two other researchers employed by the CAL—was funded by the Geography Branch of the Office of Naval Research and the Bureau of Naval Weapons, two of the major funders of projects at the CAL. The use of "automaton" in this paper and related Perceptron publications addresses the ambivalent status of computational devices and machine learning technology at the time. The earliest implementations of machine learning algorithms were custom-produced physical devices, not general-purpose digital computers. These devices or machines were conceived of as automatons in the cybernetic sense—self-organizing devices that could be shown a series of training data, learn a rule for discriminating between supplied classes, and use the rule to classify new data. CAL remained invested in the Perceptron model for several years after Rosenblatt's departure for Cornell University; they used both their hardware implementation in the Mark I and, more frequently, simulated Perceptrons on their general-purpose IBM digital computers.

In their earlier paper, Holmes and his coauthors situated their work in the explosion of this highly instrumentalized form of surveillance culture. "The extremely large volume of photographic material now being provided by reconnaissance and surveillance systems," they write, "coupled with limited but significant successes in designing machinery to recognize patterns has caused serious consideration to be given to the automation of certain portions of the photo interpretation task."[54] The goals articulated by this early investigation of nascent computer vision are not as lofty as those stated by later researchers and critics. Holmes and his colleagues wanted to find a way to reduce the "boredom and fatigue" of human photointerpreters. They recognized that the heating up of the Cold War was leading to large numbers of photographs that needed to be processed, but there were few demands for the rapid, automated processing of these images to locate objects of interest—a situation that would soon change. This change would be the sharp increase in reconnaissance and surveillance photographs taken by U.S. military and intelligence agency–operated aircraft, including Lockheed's earlier U-2 (first flight in 1955) and the SR-71 Blackbird (first flight in 1964). Simultaneously, digital computers were becoming increasingly powerful, and recent developments in imaging technology had enabled the digitization of many of these photographic images. Automating the activities of the photointerpreters would enable the military to process and analyze the visual representation of larger geographical spaces and on a more frequent basis. It was—and remains—a difficult problem to solve, but the government-funded research scientists believed that they now had the resources necessary to automate this highly systematic and urgent form of photograph interpretation.

Segmentation

Locating objects within photographs—essentially producing short descriptions of the major items within the images—is an incredibly complicated task that requires numerous interpretive decisions to be made about what constitutes a major object of interest. Searching across pixels for pattern matching is computationally expensive and introduces a greater probability of a mismatch or false-positive result. One of the largest problems identified early in the development of computer vision was the need to produce some sort of data reduction. Images taken of natural scenes are flooded with information, most of which is not of interest to the task at hand. "Natural" images are unlike those training images, which might be thought of as object portraits. The object portrait foregrounds the single object of interest and features a minimal background scene. Close, cropped images of a single object are focused. Everything within the image, almost every single pixel, provides information about the object of interest. Components present within the scene define the range of possibility for this object. Airplanes, to take an example from the origin scenes of computer vision, might have multiple jet engines or even propellers.

The process of data reduction that produces boundaries around potential objects is called segmentation. Like many developments in computer-assisted data analysis, this procedure was once a primary task and now has become a preprocessing stage, one step on the way to a more complex analysis. Twenty-first-century computer vision also depends on image segmentation and parcellation. It is important to separate the multiple objects found within an image to extract and categorize these objects. Sometimes these segmented objects are handled in parallel, enabled by developments in parallel computing; for other applications, a high-level and simplistic classification method is used to handle groups of segmented objects in different ways depending on their type. Holmes and colleagues define the segmentation problem as "determining where the pattern of interest begins and ends (as in speech recognition problems) or how one defines those precise regions or areas in a photo which constitutes the patterns of interest."[55] Developing some solution to the problem of segmenting images into smaller areas or patterns was essential to making photographic analysis computationally feasible. The segmentation problem and pattern recognition or object detection share some goals and differ mainly in terms of abstraction and scale. We can think about image segmentation as a form of pattern recognition that searches for much larger patterns. Once these larger patterns have been identified, they can be searched for smaller features, and these features can be compared to other known patterns.

For Holmes and his colleagues, photointerpretation is an iterative and recursive task that begins with a simple segmentation of image regions into two major categories: blobs and ribbons. With these two primary objects of interest, they embed into the photointerpretation task their clear understanding of the motivating military goals. These geometrical shapes are the primary mechanism by which they create and draw boundaries around objects of military interest. The two segmented object categories have a relation to each other that the procedure tacitly acknowledges in the presence of both within the frame of a single image, but the methods are not yet sophisticated enough to encode the nature of their relation. Equal weight is given to the objects identified as blobs and ribbons, although segments recognized as blobs were flagged for additional analysis. These two basic shapes are defined by simple length-to-width ratios. The photointerpreter searches particular military targets. Targets of particular interest would be the infrastructural ribbons, or "roads, rivers, railroad tracks," used for the transportation and linking of the other primary shape, blobs, which would include "aircraft storage tanks, buildings, runways."[56] Frequently collocated, the presence of multiple blobs and ribbons in an aerial photograph would indicate to photointerpreters that they were looking at a potentially important military target. Such sites would be marked by the presence of both vehicles and the infrastructure to move these vehicles.

Holmes published his "Automatic Photointerpretation and Target Location" article in 1966. While some researchers in computer vision, including Rosenblatt, abstracted their problems or attempted to provide basic science answers to the questions proposed by their home institutions, funding agencies, and military supporters, Holmes's research was at the outset directly applicable to military goals, making frequent reference to examples from the field.[57] The digitization of photographic images for both of Holmes's initial experiments required the design and manufacture of custom hardware to acquire and quantize image information from photographs. Holmes recognized that there was a need to separate an image into its basic object components in order to locate military targets. Using Rosenblatt's simplified Perceptron algorithm meant that Holmes needed to be able to automatically produce classifications into binary categories. For example, a detected image component would either be or not be classified as a particular type of airplane. Holmes's military targets were explicitly those that would interest military staff concerned with activities in Vietnam. The photointerpretation task, like almost all applications imagined within computer vision, was a search for human-made rather than natural objects. The need to find such objects provided direction for the field and produced lasting consequences. Holmes's solution to the segmentation of images was the application of preprocessing filter known as the Kolmogorov-Smirnov (K-S) filter. Holmes explains what this filter added to his search for human-made objects:

> The filter is designed to assign every resolution cell of the photo to one of a number of collections, each collection of which exhibits essentially the same amplitude distribution of grey level. The additional, very essential, constraint placed on the process is that all cells in each collection must be geometrically contiguous.[58]

This filter essentially represented each digitized photograph as a set of all possible nine-pixel or "resolution cell" (using a 3 × 3 pixel matrix) matrices. If one adjacent 3 × 3 pixel matrix shares features with another, then Holmes grouped these together into a larger unit that he calls an "agglomerated group." The K-S filter thus reduced an image into many possible component parts. Many of these would be uninteresting background objects, photographic artifacts, or noise, but some would be possible matches for a library of known objects. The K-S filter, and indeed almost all computer vision operations, works on altered pixel values. These altered values might include normalizations that reduce or expand the range of values, converting them to different scales including binary values, or averaging to blur values and to make more pronounced distinctions and objects. In so doing, computer vision deploys assumptions baked into a theory of vision as the extraction and intake of data from the visual field that have little correspondence to human perception.

The assembling of a library of different possible source images was essential to the template method of pattern detection, a pixel-based method that would eventually be extended to more complex features for categories that contain variation, such as recognizing airplanes with different configurations of equipment and eventually recognizing faces. Templates are best defined as closely cropped grids of pixel values that represent a distinct object of interest rather than an entire scene. These template objects are disembedded, usually manually, from the visual scenes in which they were originally located. Although extracted templates are closely cropped, traces of the scene from which they were removed remain—for example, within the image's perspective, possible background objects captured in remaining pixels, the template's resolution as a function of distance from the object, or the shading. Pixel-based pattern-matching methods, given a pair of digitized aerial surveillance photographs taken from a similar distance using lenses with a similar focal range, could match a template image taken from one image with the same object found in the other. Pattern-matching techniques are pixel based and thus are categorized within what I have characterized as the first ontological shift within computer vision. Within this scheme, the patterns are compared at the pixel level, which is to say that the goal becomes a search for the minimal difference of pixel values between pairs of target and source images. This scheme does not require an exact match, only a similarity of encoded representations among supplied samples. The performance can also be increased by reducing distortions and noise from differences in lighting and slight perspective shifts by converting the images and patterns to lower resolution images by using, to take one common strategy, binary values that encode only the absence or presence of an object.

The template matching method of computer vision continues to be popular because it is easy to understand and relatively simple to operate. Once defined, templates can be moved across the entire image space of the target image. Although template matching is location variant, meaning that a small template can match just pixel regions within a larger image, it remains limited in terms of matching because it requires that the template and the target image share the same orientation. This means that rotated objects or those captured from different perspectives will most likely not produce a match for the selected algorithm.

Figure 8 shows one of major target problems defining the task of photointerpretation in the 1960s. This photographic image is

Figure 8. Source image for a photointerpretation task. From W. S. Holmes, H. R. Leland, and G. E. Richmond, "Design of a Photo Interpretation Automaton," figure 4 in *Proceedings of the December 4–6, 1962, Fall Joint Computer Conference on Computers in the Space Age. AFIPS '62 (Fall)* (December 4–6, 1962, Association for Computing Machinery, New York, New York), 27–35. https://doi.org/10.1145/1461518.1461521. Permission conveyed through Copyright Clearance Center Inc.

reproduced in one of the early papers developing techniques for computer-aided photointerpretation. It is an aerial surveillance photograph that presents several problems to the researcher, including identifying the background and counting repetitions of similar objects in the foreground. Human interpreters would have no problem describing this image as containing four airplanes that have most likely recently left an airfield and are now flying over water in the same direction. In order to identify the space on the portion of the image capturing land as an airport, interpreters would need to recognize the long parallel strips of surfaced space not as roads but as runways. The major task is to identify and automatically count the number of airborne airplanes. In this instance, the task is made easier by the clear distinction between the image's background and foreground. This understanding of images and this model of vision became widely popular as researchers competed for funding from military agencies interested in applying automation to the task of viewing and interpreting regularly captured images of military and industrial facilities.

Holmes makes the observation that the prospects for reliable automatic photointerpretation are "more heavily dependent on preprocessors than on image classifiers"[59]—an important insight that remains crucial to any understanding of computer vision. Image preprocessing steps, even those that add or remove noise or reduce the amount of information available in an image by converting to gray-scale or binary images, produce radical redescriptions of the original image. The higher-level feature extraction methods that will be shortly introduced also dramatically improve some kinds of object recognition tasks. These new representations can greatly improve the results of classification, but potentially at some cost in terms of computational complexity, the particularities of the images, and the individual differences found in the represented objects.

The Rise and Fall of the Perceptron

Frank Rosenblatt staged his first public demonstration of the Perceptron for reporters on July 7, 1958. The framing of this event and his invention's reception would become one of the most important determining factors for the future of Rosenblatt's research. The event also established the template for what we might call the hype cycle of machine learning and artificial intelligence technology. Rosenblatt was only able to demonstrate the Perceptron algorithm, not the entire Mark I system, which had not yet been constructed. This was an emulation of a Perceptron, the initial software implementation, running on the industry standard computer of the time, an IBM 704 at the U.S. weather bureau in Suitland, Maryland. Although it was not an expectation at the time, this emulated form of the Perceptron running on a general-purpose digital computer would soon become the primary way it was used. Neural network machines were not able to replace digital computers for several reasons. Among the most important were their high cost of development and their limited applications. These factors were to appear again and again in the history of neural network–inspired alternatives to general-purpose computers, even when these devices were constructed from large networks of smaller digital computers.[60]

On his way to Maryland, Rosenblatt stopped at the offices of the *New Yorker,* which would run its own story about Rosenblatt and the Perceptron. Under the title "Rival," the *New Yorker* positioned Rosenblatt's to-be-demonstrated Perceptron as a potential rival to not only the IBM 704 computer that presently simulated the Perceptron but also the human brain:

> The distinctive characteristic of the perceptron is that it interacts with its environment, forming concepts that have not already been made ready for it by a human agent. Biologists claim that only biological systems see, feel, and think, but the perceptron behaves *as if* it saw, felt, and thought. Both computers and perceptrons have so-called memories; in the latter, however, the memory isn't a mere storehouse of deliberately selected and accumulated facts but a free, indeterminate area of association units, connecting, as nearly as possible at random, a sensory input, or eye, with a very large number of response units.[61]

The *New Yorker,* like other news outlets, focused on the yet-to-be-developed visual input system that they imagined would seamlessly interact with the Perceptron's hardware to give the impression of this fantasy automaton. Digital computers, in reality and in the public imagination, were loud and large at this time. The complex input procedures used to enter data and the privileged access to computers made the fantasy articulated by the *New Yorker* of an automaton sensing and interacting with its environment all the more compelling. The reality was still primarily punch card input and paper-printed output.

After this interview, Rosenblatt continued on his way to Maryland. His demonstration of the Perceptron did not resemble the fantasy scene produced in New York. There was no sensory input and no digital eye. There was no camera; nor would a photoreceptor array eventually be developed. Data were supplied in the form of simple punch cards. The event itself was widely reported in the U.S. press. The handling of this demonstration, by both Rosenblatt and the reporting press, might be said to have produced the first great expectations in the boom-and-bust story of artificial intelligence. The *Daily Boston Globe* published their story, sourced from an AP reporter, under the title "Shades of Frankenstein! Navy's Going to Build Robot That Can Think." The *New York Times* also engaged in hyperbolic descriptions of the possibilities—"The Navy revealed the embryo of an electronic computer today that it expects will be able to walk, talk, see, write, reproduce itself and be conscious of its existence"—but provided a more concise and staid summary of the actual demonstration:

> In today's demonstration, the "704" was fed two cards, one with squares marked on the left side and on the other with squares on the right side. . . . In the first fifty trials, the machine made no distinction between them. It then started registering a "Q" for the left squares and "O" for the right squares. Dr. Rosenblatt said that he could explain why the machine learned only in highly technical terms. But he said the computer had undergone a "self-induced change in the wiring diagram."[62]

These demonstrations of the Perceptron algorithm, despite the trivial nature of the learning demonstrated, captivated the journalists in attendance and produced a sensational response in their reporting, thus raising stakes of future developments in machine learning and establishing a pattern for such technologies of making pronouncements and promises that they could not deliver.

In the early 1960s, Marvin Minsky, one of the Dartmouth artificial intelligence conference attendees, started work organizing a new research group on artificial intelligence in collaboration with John McCarthy. McCarthy had left Dartmouth for MIT shortly after the conference. In 1961, Minsky was commissioned by the editors of the *Proceedings of the IRE,* the major publication of the Institute of Radio Engineers, which was later to merge with the American Institute of Electrical Engineers to create the Institute of Electrical and Electronics Engineers, or IEEE, to produce an important field survey of machine learning and artificial intelligence. Minsky's essay, titled "Steps toward Artificial Intelligence," would set the priorities, goals, and self-understanding of the field for many years to come.[63] Minsky, like many others, turned to visual pattern recognition, what would later come to be known as computer vision, to provide an example of real-world applications for the algorithms and nascent systems that he was describing. Rosenblatt's photoperceptron model, with its conjoined image acquisition technology, provided Minsky with an ideal object to critique. Taken together, the Perceptron was proposed as a visual pattern-recognition system. Minsky found two major faults in Rosenblatt's model. First, he did not believe that the randomized connections between elements found in Rosenblatt's design could lead to the number of properties required for classification of complex patterns. Second—and this problem was directly connected to the use of the Perceptron for primitive computer vision tasks—Minsky was concerned that the lack of automatic preprocessing available for template matching, a key technique in visual pattern recognition, would require extensive manual adjustments of images before classification:

> It does seem clear that a maximum-likelihood type of analysis of the output of the property functions can be handled by such nets. But these nets, with their simple, randomly generated, connections can probably never achieve recognition of such patterns as "the class of figures having two separated parts," and they cannot even achieve the effect of template recognition without size and position normalization (unless sample figures have been presented previously in essentially all sizes and positions). For the chances are extremely small of finding, by random methods, enough properties usefully correlated with patterns appreciably more abstract than those of the prototype-derived kind. And these networks can really only separate out (by weighting) information in the individual input properties; they cannot extract further information present in nonadditive

> form. The "perceptron" class of machines have facilities neither for obtaining better-than-chance properties nor for assembling better-than-additive combinations of those it gets from random construction.[64]

Minsky here makes several critiques of Rosenblatt's model and what he sees as its typical application. He does so in order to argue that the Perceptron is not up to the task at hand (that is, pattern recognition), and further that it would not be able to produce statistically significant classification results from its use of randomly generated connections between the units or neurons in the network. Minsky was hardly unknown to Rosenblatt; indeed, they had gone to the Bronx High School of Science together and appeared together at conferences.[65] In critiquing the Perceptron, Minsky brought renewed attention to the algorithm and kick-started a debate that would continue for almost a decade.

After the initial enthusiasm for machine learning and artificial intelligence, the tide began to turn against these methods. Frank Rosenblatt's *Principles of Neurodynamics: Perceptrons and the Theory of Brain Mechanisms* (1962) opens with a defensive preface responding to what Rosenblatt sees as a hostile reception to his invention. Part of this hostility, Rosenblatt suggests, is the result of a misunderstanding of his device's purpose, which produces a different explanatory frame than the one he desires. Calling Minksy's "Steps toward Artificial Intelligence" the "entertaining statement of the views of the loyal opposition," Rosenblatt reframes the Perceptron not as a general system for visual pattern recognition but as research model, a brain simulation:

> First, was the admitted lack of mathematical rigor in preliminary reports. Second, was the handling of the first public announcement of the program in 1958 by the popular press, which fell to the task with all of the exuberance and sense of discretion of a pack of happy bloodhounds. Such headlines as "Frankenstein Monster Designed by Navy Robot that Thinks" (Tulsa, Oklahoma *Times*) were hardly designed to inspire scientific confidence. Third, and perhaps most significant, there has been a failure to comprehend the difference in motivation between the perceptron program and the various engineering projects concerned with automatic pattern recognition,

> "artificial intelligence," and advanced computers. For this writer, the perceptron program is *not* primarily concerned with the invention of devices for "artificial intelligence," but rather with investigating the physical structures and neurodynamic principles which underlie "natural intelligence." A perceptron is first and foremost a brain model, not an invention for pattern recognition.[66]

Rosenblatt's attempt to draw distinctions between "artificial intelligence" (note his use of quotation marks to surround this term) and the Perceptron comes out of his belief that the public as well as the major U.S. funding agencies had become disappointed with his machine. The cultural discourse networks, newspaper reports, and scientific communications are pulled into Rosenblatt's response because the rhetoric and explanatory frames are incredibly important to the politics of technology. They also have material costs in terms of the perception of funding agencies and the public's willingness to support large-scale technological projects. We can read Rosenblatt's response as a counterdiscourse, an attempt to clarify the meaning and purpose of his project against the dominant explanatory frames that have been imposed on him and his research.

Minsky would double down on his critique of what he saw as the limits of the Perceptron model with his coauthor Seymour Papert in their 1969 book *Perceptrons.* This book has been widely understood as one of the major causes of the changes in the U.S. Department of Defense's funding priorities for artificial intelligence and machine learning, as well as the general cooling of excitement over the technology that led to what many called the "AI winter." The arguments made in *Perceptrons* essentially divided machine learning from artificial intelligence, and the authors sought to purify AI research by excluding machine learning approaches from the field. Minsky and Papert critiqued Rosenblatt's basic Perceptron model, which they took as the exemplar of contemporary neural networks, for its inability to execute an XOR, or "exclusive or" function.[67] Implementing the XOR function would necessitate additional complexity in the form of a multilayer neural network. The linear classification model implemented by a single-layer Perceptron was unable to handle the case in which presented data might be classified by both desired categories. In Léon Bottou's foreword to the 2017 edition of Minsky and Papert's *Perceptrons,*

Bottou writes: "When the first edition of *Perceptrons* appeared . . . perceptron research had not yet produced a successful application in the real world. In fact, the most obvious application of the perceptron, computer vision, demands computing capabilities that far exceed what could have been achieved with the technology of the 1960s."[68] While there was plenty of ongoing research into the workings of multilayer neural networks and other solutions to the problems identified by Minsky and Papert, the discourse around machine learning turned negative, and funding priorities would shift in the coming years. The discourses and counterdiscourses invoked above were mostly technical in nature; they concerned either mathematical and logical limitations, or different understandings of the problem for which the Perceptron was offered as a solution. There were other critically important counterdiscourses on the part of change agents that would soon challenge the domination of the military-industrial complex in universities and contribute to these shifts in funding priorities.

Rosenblatt died young, on July 11, 1971, his forty-third birthday. He died in a tragic accident on his sailboat, *Shearwater,* while crossing the Chesapeake Bay from Oxford to Annapolis, Maryland.[69] Rosenblatt was accompanied by two young friends. The obituaries that appeared shortly after his death stated that he was "swept overboard" and drowned.[70] In his remarks at a Cornell University memorial, Richard O'Brien, head of Cornell's division of biological sciences, called Rosenblatt a "victim of the Mansfield amendment" and noted that while he once had access to hundreds of thousands of dollars in funding for his research, "within a few years that money melted like summer snow and soon he had very little left in the last few months."[71] Nonetheless, Rosenblatt had started up numerous other scientific investigations. At the time of his death, Rosenblatt was engaged in several new projects; one of these brought him from mechanical back to biological models and returned him to his earlier work as a behavioralist. Rosenblatt was interested in the possibilities of moving or transplanting learned behavior from the brain of one organism to another. Instead of his conception of storing responses within a digital simulation of neurons, Rosenblatt now attempted to "migrate" stored responses from one rat brain to another by injecting extracted brain tissue into a live rat.[72] Another major project, which his obituary in the *Washington Post* stated was occupying half his time, was his "working on a study of air war in Indochina for the Cornell Center for International Studies." This research effort was dedicated to exposing the extent of the violence and the human, political, and ecological costs of the ongoing war. It also became a turning point in Rosenblatt's career by joining his progressive politics to his research in perception and computer vision that had been used for so many military projects and activities. Rosenblatt's work with the Air War Study Group would result in the publication in October 1971 of a preliminary report titled "The Air War in Indochina." It would be extended and published in 1972 in book form by Beacon Press and dedicated to the memory of Frank Rosenblatt.[73] In 1972, Marvin Minsky and Seymour A. Papert produced a second printing of Perceptrons and added to their text numerous handwritten annotations and notes, including a new dedication reading, "In memory of Frank Rosenblatt," thus concluding the structuring binary that had pitted Rosenblatt against his loyal opposition.

« 3 »

Describing Pictures

From Image Understanding to Computer Vision

Image recognition tasks including face and object detection are typically performed with ease by the majority of adult human beings, but such tasks have proved exceedingly difficult to formalize into a set of repeatable steps and procedures suitable for a computer. These tasks, however, have been approached with much enthusiasm and hubris throughout the history of computer vision. With some early successful demonstrations that worked on highly simplified images belonging to a reduced set of possible categories, researchers set their sights on more complicated problems. During the 1960s and 1970s, machine learning and pattern-recognition research quickly shifted focus from basic problems in computer-assisted perception dominated by pixel-based data to scene analysis and complex pattern recognition using high-level symbolic representations. This shift, as has already been suggested, involved several different forms of expansion. The binary logic of a forced choice between two defined categories that was embedded in the major models developed during the Cold War period of the late 1950s and early 1960s, including in many demonstrations of Frank Rosenblatt's Perceptron, eventually gave way to research directed toward the identification of multiple classes of known objects. Those developing pattern-recognition techniques assumed that they knew what could be found in images; collections of known objects were assembled, described, and placed into databases. Researchers, driven primarily by goals and funding derived from the United States military for the war in Vietnam, were focused on specific image tasks that they thought had solutions on the near horizon. These solutions were destined to become a crucial part of the "electronic battlefield."[1]

The field now known as computer vision was born in response to the difficulties discovered during this second phase of pattern recognition, when the ontology of the image was once more altered. The task of locating complex human-made objects embedded in an unfamiliar environment proved more difficult than initially predicted. The various research projects formed as a response to requests for proposals from U.S. military funding sources during the 1960s and 1970s were initially positioned under the rubric of image understanding. When the field was later reconfigured with the ambitious title of "computer vision," it had to undergo a transformation. As part of this transformation, it was acknowledged that the vision directing the future of the field's research would not emerge solely from within computer science but would be produced as the result of interdisciplinary and collaborative work that combined the resources of several distinct disciplines. Knowledge of computational and statistical methods was no longer enough to solve these complex problems. Domain-specific knowledge of a variety of kinds—visual arts, linguistics, information theory, and even philosophy—needed to be brought together to establish an ontological framework for computer perception and understanding.

The recognition of discrete objects within photographs proved to be computationally difficult. One of the major critiques of Frank Rosenblatt's Perceptron algorithm and his hardware implementation in the form of the Mark I Perceptron was that the algorithm required uniform and standardized images to be used as input data. These input images had to be closely cropped; the depicted objects needed to be framed exactly the same way, undergoing what artificial intelligence researcher Marvin Minsky calls "size and position normalization" before their input as quantized data for the Perceptron algorithm.[2] This normalizing procedure enabled early computer vision techniques to match patterns at the level of the entire input image; any distinction between foreground and background was lost, and the image itself as a grid of values became the pattern. Calculations of similarity among the images were thus made at each of the elements within the digitized image. This was, *avant la lettre,* a "pixel-wise" comparison and pattern matching. (The term "pixel" for a picture element was first used in 1965 and was a contested term of art until the 1970s.[3]) As Minsky demonstrated, this was a major limitation of early machine learning methods in their application to visual images. The Perceptron performed pattern matching across the entire image; the image itself was treated as the source and target pattern rather than objects found within the image. The recognition of complex objects embedded in larger scenes, rather than pattern-matching tasks applied to well-established sets of printed characters or closely cropped images of faces, required much more complicated methods that could operate across different scales and orientations. This period began with as much excitement, promise, and hype as the era of intelligent machines inaugurated by public demonstrations of the Perceptron. But much like the problems plaguing the development of the Perceptron and the larger field of machine learning recounted in chapter 2, the roadblocks encountered during the development of the area of research known as image understanding demonstrates the difficulties involved in the extraction of meaningful information from a closely bounded and tightly framed image.

By tightly framed, I mean to say that these early approaches to matching and recognition of digitized images, even those images with expansive backgrounds and settings, were constrained by a special type of aperture that acts like a virtual filter to render image data the only source information about the contents of the image. The history of the area of research known as image understanding is organized through a reckoning with the limitations of this framing. This history begins with the initially hopeful efforts to break down a complex problem into smaller, more computationally manageable problems. The efforts were stalled by both technical and theoretical limitations. Faced with an inability to deliver on the promises made to the press and funding organizations, researchers realized that automated analysis of image data could not be achieved by staying within their original framing of the problem; instead, they required supplemental information and knowledge of the world. Much of this additional information, they discovered, was entirely external to the content of their digitized photographs—that is, to the data. The gradual incorporation and formalization of this knowledge about images, rather than data derived from images themselves, required a reconstruction of the original problem and a recentering of the role of human perception, evaluation, and decision-making in the interpretation of photographic images.

The methods and algorithms developed during the photo-interpretation phase of computer vision research in the 1960s and 1970s are crucial not only because these were the dominant methods of the period but also because a number of these algorithms remain in popular use today. Additionally, the general conception of the image and the ontological understanding of an image as a collection of features developed at this time established the central model that would be used for decades to come. Most of the military problems that these algorithms were designed to solve resided within contemporary applications that made use of these ways of understanding images. Even when the algorithm has been modified or replaced with more advanced or complicated procedures, the historicity of these earlier algorithms and methods can be located as a body of accreted knowledge of the prior executions of these algorithms, which is to say that prior framings and conceptions are activated by every new iteration of these methods. The competing models of machine learning algorithms are one such example, with subsequent generations borrowing from innovations made by competitive models in the previous generation. Algorithms are discursive and thus inheritors of a network of dominant discourses and counterdiscourses. Another aspect of this historicity is to be found in the norms embedded within the limited range of collected historical visual data. Machine learning algorithms learn rules and criteria from supplied data sets, and the diversity (or lack of diversity) reflected in these data sets produces and reproduces the norms.[4] Those norms established through methods such as averaging, with the average itself being normative in form, continue to project their influence beyond their moment of collection or inception. As Joy Buolamwini and Timnit Gebru demonstrate, the presence of multiple categories of bias, gender, and race in the case of the pretrained commercial face recognition classifiers they analyzed makes an intersectional form of analysis a crucial tool in exposing these norms.[5] Although some computer vision techniques arising from these photointerpretation tasks are unavoidably linked to military imperialism and nefarious surveillance technologies, other techniques have slightly more tacit connections to these tasks but nonetheless embed within them normative assumptions about humans and the world.[6]

Picture Descriptions

Conceptually, all digital representations of image data are symbolic descriptions of sensed scenes. Computer vision researchers increasingly would use the term "sensed" or "seen" rather than "perceived," as perception requires the ability to understand the relations found in the content of an image. Sensed is the acquisition of sensory input detected by a camera and its lens. The pixel intensity values found in an image taken with a digital camera are sensed and then produced by sampling light from a sensor grid. These values are approximations of the reflected light acquired by sensors. The sensed and quantized data can thus be said to provide a symbolic representation, in this case easily numerically manipulable, of the sensed scene. Likewise, similar sensors can scan previously acquired photographic images, which have already undergone some level of filtering and framing. Although one might differentiate, for example, between higher- and lower-resolution versions of the same image, both data objects provide symbolic representations of the same sensed scene. The consistent extraction of key, representative, and comparable pictorial data from these grids of pixel values proved and continues to prove to be a difficult task. Determining what reduced set of features can be detected and extracted from an image—and reducing is necessary because one cannot count every individual pixel in an image as meaningful—is complicated because of the large number of pixels, especially as technology develops to capture higher-resolution images. Two images of the same scene taken from different perspectives, at different times of the day, or even with two different cameras will have a large number of changed pixels, even within a small, focused window. Image descriptions are essentially the same as feature detection: standardized, reliable, and repeatable representations of the content of an image. Without reliable computational methods to detect and extract representative features directly from images, computer scientists and engineers needed to search elsewhere for features that could serve as input for their machine learning algorithms.

Early in the history of computer vision, it was necessary to make a distinction between pictorial, which is to say pixel-based image data, and higher-level symbolic data. Symbolic data, at least for some early researchers, might be best thought of as metadata rather than data. If "data" is the word used to refer to sensed image data, then "metadata" would be data about these data. In the case of computer vision, this symbolic metadata would describe the content of the image rather than comprise content-derived representational data. Yet when attached to automatically recognizable features within images, symbolic data would eventually come to be understood as representational rather than descriptive. This change in understanding had important ontological consequences for the theory of the image within computer vision. Computer vision turned to such symbolic descriptive data in two forms: first, as expert knowledge about images and the objects represented in them in grammatical or textual form that could be encoded, processed, and used alongside digitized images to enhance knowledge extraction and retrieval tasks; and second, as higher-level descriptive features that could be used to identify a class of complex objects that would be represented in pictorial data in different ways but that share the same symbolic representation.

In the twenty-first century, there are a number of computer vision algorithms that are best framed as descriptive algorithms, and the pattern-matching methods described in this chapter continue to be important tools for computer vision applications.[7] Because training image data are typically rendered with labels and these labels tend to be simplified, discrete category descriptions of the image; for example, for the label "tree," the accuracy of an algorithm is assessed by the correct labeling of an image with its descriptive label. We might think of this as a captioning task. Many contemporary web-based or social media applications, for example, make use of automatically generated text-based descriptions of images for the purposes of "alt text" tags for accessibility purposes. These are essentially descriptive captions. The Microsoft Office suite also has bundled machine learning–powered computer vision tools that recognize images and objects within images and automatically annotate images with these descriptions.[8] The algorithm captions an input image with the best-fitting descriptive label for the major content of the depicted object. In the case of hierarchical categories, that caption might include probability or confidence scores for major and minor categories. For instance, an image labeled "sugar maple" might be returned with 85 percent confidence and the parent category "tree" with 98 percent confidence.

The notion of captioning, or the idea that we might be able to link a shared referent with both a visual and textual representation of an object, animates both contemporary and historical computer vision. In matching object with concept, computer vision embeds ontologically normative categories and judgments as transcendent and unquestionable. Indexing and cataloging concepts with text have been thought of as easier than doing the same with a set of images. From a theoretical standpoint, we might consider captioning and describing images with text—that is, with language—analogous to their symbolic representation as clusters of image cells or even a matrix of pixel intensity values. In transposing the digital representation of an image into another representation space, an algorithm provides a symbolic representation of that image in relation to other images. In Yuk Hui's account of the recurrent causality of digital objects, as introduced in chapter 1, the initial image—what we could call the first level of digital representation—presupposes the relations and connections of subsequent algorithmic transformations. Procedures to break down complex objects and eventually entire visual scenes into hierarchical object relations has its origins in 1970s-era research on picture grammar and pictorial structures. The same is true of many commonly used pattern-matching techniques in computer vision today. The grammar and structures were used to formulate the basics of image descriptions. These structures and grammars provided a way for computer scientists to increase the accuracy of algorithms by identifying common criteria that would bundle together a set of image features belonging to a larger object class.

Martin A. Fischler, a key early computer vision researcher, began work on this problem as early as 1969, before he moved to SRI and while employed by the Lockheed Corporation's Palo Alto research laboratory. Fischler, like others involved in early military-funded computer vision research, was an army veteran. In a paper delivered at the first International Joint Conference on Artificial Intelligence, which took place in Washington, D.C., in 1969, Fischler described the impasse reached with research into the extraction of meaningful pixel-level features for image classification: "The problem of obtaining a suitable representation for a picture is much more difficult than the problem of classifying the picture once a satisfactory representation is obtained."[9] The representation type that interested Fischler and others was a set of features computationally derived from image data. Although accurate classification images, given a reliable set of features, was possible, it was more difficult to determine a procedure to reliably extract these important features across numerous images. The identification and extraction of features from images remains a challenging problem today. Fischler wanted to find an alternative method of feature detection that could be used as the basis for existing classification algorithms. Fischler began his research into image descriptions with the assumption that people, as expert image detectors, might be able to describe an object in language in a manner that would enable another person to understand and then recognize that object, even if it was embedded in a larger scene.

In his article, Fischler posits an important distinction between seeing and perceiving—a distinction common in the discourse of the emergent field of computer science, and in particular discussions of artificial intelligence.[10] This distinction enabled him to split the work of humans and machines in his conception of computer vision. Fischler writes:

> We first note that there is a sharp distinction between "seeing" and "perceiving." "Seeing" is the passive reception of visual data; we might say that a TV camera "sees" the object it is focused on. "Perceiving" is an active creative process in which a visual scene is decomposed into meaningful units. This decomposition is a function of the stimulus pattern, the vocabulary and experience of the observer, and the "psychological set" of the observer.[11]

For Fischler, such a split between seeing and perception would need to leverage human knowledge about visual objects in order to make use of the machine learning methods that were created alongside some of the more fundamental computer vision techniques. Because these machine learning techniques built on already established statistical methods, such as the nearest neighbor rule, the k-nearest neighbors (k-NN) algorithm, versions of linear discrimination, and Bayes's rule, they were much more advanced and capable than most of the existing feature detection methods for even simple two-dimensional visual objects.

Fischler's solution, the LSADP (line segment analysis and description program), was able to measure simple line drawings by following the contours of lines drawn with a felt-tip pen on 35mm film. Using a flying spot scanner, Fischler and colleagues instructed the program to start searching for a black area. Once one was located, the program would try to follow contours by searching for neighboring black areas. The "image" would be represented in simple binary values recording either the absence or presence of black. Their method generated this image representation as numerical records of what they referred to as "hits" in the form of lines detected. Human-legible linguistic descriptions of these values could then be generated from these data. Fischler describes this translation from numerical values to linguistic descriptions as follows:

> The linguistic description generated in the LSADP has some interesting characteristics. There are currently some 500 words and phrases which form the generator's vocabulary. As the line segment analysis proceeds, the linguistic phrases are assembled in a buffer area. Numeric values are converted into appropriate descriptive terms by means of translation tables. The specific table employed to translate a numeric quantity can vary depending upon the preceding analysis. Thus, for example, a line segment which forms the boundary of an enclosure must have approximately twice the arc length needed by a straight line to be given the attribute value "long." If there is too much ambiguity associated with a measurement, the corresponding descriptive phrase may be completely deleted, and under some conditions, a phrase previously entered into the buffer area may also be deleted.[12]

This method provided Fischler with the basics of a method to generate linguistic descriptions of visual data. These descriptions were convoluted and produced from the terms of human perception (that is, a line described as "long" would only be meaningful to a perceptive human), and they were generated from simple line drawings, but this gave him some insight into the possibility of the "machine description of graphical data." A partial program-generated description of a curve is described as such: "The diameter of this curve is coincident with the line between the terminal points and has a left diagonal orientation. The curve is open to the left. The approach to vertex 81 is from below and to the right."[13] The LSADP performed pattern matching on textual and numerical values; it performed searches across tables of both kinds of data. Matching descriptions that were based on human-supplied text, especially when this text was codified into established categories, was not difficult, and methods were available to deal with these kinds of data.[14]

For the time being, any machine vision of images would need to be supplemented with trained and expert human perception of objects in order to produce descriptions that could eventually be linked together, then organized into hierarchical structures, to describe entire scenes represented by pictures. Before working on these higher-level scene descriptions, computer scientists needed reliable descriptions of individual objects or segments. Once generated, these descriptions could be processed by an algorithm for classification or put together into longer sequences. This task was understood by researchers to be similar to the problem that confronted those working in linguistics on what would become natural language processing for text: how do you break down complex but related objects into a set of smaller problems that can lead to an account of the whole? The reduction of an image into its component or constitutive parts enabled researchers to associate these individual component objects with descriptive labels that provided what they termed a "grammatical" description. The grammar-based approach allowed for the construction of an image or scene from these smaller grammatical components. Researchers turned to language as a resource for classifying images because they recognized that the complete automation of military photointerpretation was not going to be computationally feasible, at least in the short term. What the digitized photographic data and the natural language descriptions of photographs or segments of photographs shared was their status as symbolic representations of the content of the image. These representations, researchers proposed, were different methods of information storage and retrieval. If descriptions of images could be formalized, if there could be some consistency of the descriptions, then this part of the photointerpretation task could be successfully automated by borrowing procedures from existing research.

Fischler returned to this work with a Lockheed colleague, Oscar Firschein. Building on Fischler's prior idea to establish a referential database of human-understanding textual descriptions of visual objects, Fischler and Firschein constructed a plan for a more robust archive of descriptions that would draw not only on the resources of human perception with its sense of relative positioning and relations between objects but also the archival and retrieval of scene-specific knowledge. Their 1971 article, which acknowledges funding by Lockheed as well as the Office of Naval Research, surveys and evaluates several grammar-based, descriptor-based, and procedure-based methods of retrieving and describing images. Like most early and foundational work in computer vision, Firschein and Fischler's research was explicitly designed to respond to particular military objectives and needs. "At present," they write, "we are conducing research in the preparation of descriptions of aerial photographs by human subjects." They proposed an abstract workflow for an image retrieval system based on queries of descriptive language of objects. This workflow is organized around the goal of understanding the objects present in a series of aerial photographs and the relations among these objects. Their system would be used to digitize photographs of scenes and store these image data alongside human-created descriptions of these objects and scenes. The separation of image data and descriptions functioned much like the image–caption relation or the data–metadata schema used in many digital systems today. Firschein and Fischler's system would enable users to retrieve matching pairs of images and descriptions from their respective archives with text-based queries of the information found in the stored descriptions.

In order to produce these robust descriptions, they understood that they would require the resources and knowledge of many disciplines.[15] Drawing on insights from linguistics, psychology, library science, and, surprisingly, creative writing, Firschein and Fischler proposed as a solution to the photointerpretation problem a complex system that enabled computer-aided photointerpreters to describe, store, recall, and match previously stored image descriptions. The project at hand was to determine:

> How human subjects approach the problem of describing an aerial photograph.
>
> The influence of instructions on the type of descriptions prepared.
>
> The importance of training and background in preparing adequate descriptions.
>
> The global structures and vocabulary used in these descriptions.
>
> The types of reasoning and the prerequisite information needed to produce descriptive statements of a deductive nature.[16]

Firschein and Fischler thought that natural language descriptions would be particularly useful for these tasks and pointed to the language of literary realism and Mark Twain as an exemplar of a description that is not "overloaded with excessive details and specifics."[17] Building on their prior work on pictorial structures with language-based descriptions, they added to the task a way to supplement the photointerpreter's memory by activating prior knowledge of what constituted the range of possibilities for any particular object.

Firschein and Fischler ended their survey of these grammar-based methods for describing visual objects and scenes with the proposal that what they termed "picture primitives" would be the optimal solution to the problems they had identified. Primitives were a set of common building blocks used to construct more complex modeled objects. They produced a distinction between two types of picture primitives, descriptor primitives and shape primitives, that enabled them to move forward with the cataloging of image data:

> Formal methods of picture description rely on the use of picture "primitives," basic entities that can be combined in a symbolic representation of a picture. In the case of descriptor-based description, the primitives can be considered as being the individual descriptor words or phrases; in the (current) grammar- and procedure-based approaches, the primitives are usually limited to denoting simple geometric shapes. Relations involving the primitives are separately defined items.
>
> It is interesting to note that the descriptor primitives are "high level," i.e. strong in semantic meaning and usually not joined by formally defined relationships or hierarchical structure, . . . while the shape primitives employed in current picture languages tend to be "low level," referring to segments of objects, and are tied together in rather complicated relationships which then comprise the higher level entities related to real world objects.[18]

Descriptive primitives were a major step forward in computer vision. The numerical values of simple line figures could now be replaced with symbolic representations of more complex shapes. What needed to happen next was the development of a method by which one might develop systems to recognize the structures of objects with well-established structures but with much internal variation. If these methods would have any application for the military goals that were driving them forward, simple objects like boxes and cylinders would need to be replaced with descriptors that could work with the variation found in what were termed "natural" scenes. Military targets, those sites containing vehicles and weapons, formed the basis of these prototypical natural scenes.

Pictorial Structures

These primarily military-funded researchers were attempting to develop computer vision systems that could locate military targets of interest by automating the work of what was then the specialized job of a photointerpreter. This work was started by William S. Holmes, who, as recounted in chapter 2, combined Frank Rosenblatt's theory and the hardware implementation of the Perceptron learning algorithm with newly developed image preprocessing techniques in order to apply these methods to the photointerpretation task. Military photointerpreters during the Vietnam War manually reviewed thousands of photographs taken with aerial photography, searching for changes to familiar sites over time, especially the appearance of new objects. It was estimated that by 1966, tens of thousands of photographs were taken every day by everything from radio-operated drones to high-powered reconnaissance jets.[19] Part of this job required the classification of different geographical regions and spaces. The solution to this problem turned out to be what was called "pictorial structures." Pictorial structures were a step forward from prior work in that these structures were no longer grammatical generalizations of common geometrical shapes but rather abstractions that could be linked together in loosely coupled templates. Pictorial structures could be used to create a template, and target images could then be searched for pixels matching this template. The template descriptive features were loosely connected; the article introducing the method likened the variable connections or distances between features—say pixels belonging to a nose and eyes—to springs.

Fischler was soon joined in his research on pictorial structures by a much junior Lockheed colleague, Robert A. Elschlager. Elschlager had recently finished an MS in mathematics from the University of California, Berkeley, and joined Lockheed Missiles and Space Company, a major division of the Lockheed Corporation. Fischler and Elschlager carried forward the nascent field of computer vision through the continued breakdown of what we might term the "image space" captured by film photography into smaller and smaller component pieces. It was quickly becoming apparent that computer-aided object identification could not operate on the scale of the entire image and that new methods of pattern recognition were required. The relatively unformed blobs and ribbons of picture elements found in William S. Holmes's approach to object recognition and described in chapter 2 were now given a sense of greater structure and relation to other potentially meaningful objects through Fischler and Elschlager's more focused approach to object recognition. If blobs and ribbons were essentially filled-in templates of large patterns of complete and individual visual objects, then these new methods would break down the represented object into smaller component pieces that made up a connected structure. This same technique would also change the understanding of the image in computer vision; an entire scene could now be understood as a complex connected structure of individual component parts. Previously, methods had looked for the presence of multiple rough proxies for objects, like blobs and ribbons. Now there would be an understanding of objects as symbolic representations made up of other symbolic representations. This new form of pattern matching, then, would involve the identification of collocated basic element parts that made up more complex objects and would enable researchers to develop more advanced object recognition techniques for tasks ranging from face detection to automatically classifying large geographic regions. The genealogy of computer vision from its birth to the present is an ongoing mutation of metaphors like these symbolic representations.

Fischler and Elschlager's approach to pattern matching, like many of the major interventions in computer vision, depends on the breaking of unity of the digitized image through the addition of external knowledge of what constitutes interesting patterns. They construct a more flexible method of detecting groupings of distinct component parts within the digital representation of a photographic image or scene. Their concept of pictorial structures as image primitives was introduced in a 1973 paper titled "The Representation and Matching of Pictorial Structures."[20] The basics of the approach outlined there remain at the core of some of the most fundamental methods within computer vision. In simple terms, Fischler and Elschlager describe a method of matching existing patterns from a source or reference image, as a description, to target image data as a solution to the problem of object detection. "We offer," they write, "a combined descriptive scheme and decision metric which is general, intuitively satisfying, and which has led to promising experimental results. We also present an algorithm which takes the above descriptions, together with a matrix representing the intensities of the actual photograph, and then finds the described objects in the matrix."[21] Realizing the complexities of photointerpretation, the version of computer vision proposed by Fischler and Elschlager would remain primarily pattern recognition and matching, although with the patterns becoming more loosely defined.

Fischler and Elschlager's method became the building block of the turn away from the dream of fully automated photo-interpretation to what was then termed "scene analysis." Their paper demonstrates the application of their technique to two major visual surveillance problems that share some interesting features: the detection and matching of face images, and the classification of terrain and objects within aerial photography. The automated analysis of "sensed scenes" was an interesting and hot topic in the early 1970s, both because this was a computationally difficult problem and because there was ample Department of Defense funding available as a result of the ongoing involvement of the United States in Vietnam. The sensed scenes that formed the core of the research in the field were digitized aerial photographic images of "natural" scenes. The recognition that researchers were addressing sensed scenes was required to move from the lofty goals of photointerpretation to the more nuanced and multisourced image understanding project. Fischler and Elschlager describe the applications of their pictorial structures method as follows: "Scene analysis and description, map matching for navigation and guidance, optical tracking, stereo compilation, and image change detection."[22] Once again, military goals defined the scope of possible applications for computer vision technology.

Fischler and Elschlager introduce methods that built on the existence of prior assumptions of what made up a complex object. It was no longer thought feasible to maintain libraries of potential patterns or templates for matching objects, especially objects likely to come in slightly different perspectives and configurations or in natural rather than experimental and controlled scenes. The images they invoke in their paper are a series of aerial surveillance images of military sites. The identification of a military as opposed to nonmilitary site would take place through the presence of certain objects. Their paper builds on the assumption that the best way to identify potentially important military sites would be through the copresence of various objects like airplanes, runways, tanks, ships, and railroads. In their formulation, these sites and their constituent objects would be subject to a considerable number of changes and different configurations. Allowing for some flexible configurations and distances between these possible objects, the authors believe, would cover a large range of configurations, although some required features must be present for the recognition of particular classes of objects. Pictorial structures are taken up in service of computer vision's ongoing concern with producing models that are representations of reality. "Since we cannot manipulate the real world object (itself)," they argue, "we attempt to construct a representation (or model) which can be used in place of the actual object."[23] While image data in the form of grids of pixel values provided a representation of a scene or object, depending on the framing, these representations were inadequate—especially in terms of the computational capacities of the time—because the pixel values simply contained too much unorganized information. Fischler and Elschlager determined that the representations would need much greater organization if computer vision was going to be able to address complex objects: "Much of the information contained in a representation will be implicit rather than explicit in form. The ability to manipulate easily the representation to extract required information is essential."[24]

Normative and Average Objects

While the template method of pattern recognition can be used to find either exact matches of the supplied template in an image or small variations, this approach does not require any conception of a class of objects. A template image featuring a single apple might be successfully matched with any number of different apple cultivars, and matching methods with lower thresholds may even match with oranges, but no sense of what makes up the category rather than the example "apple" is contained within this method. A single apple stands in as the representative of the category or class "apple." The use of a single sample distinguishes the template matching method from the approach used in contemporary neural network models, in which many sample images of different apples would be used to define the class. The creation of pictorial structures provided researchers with an additional layer of abstraction, one removed from sets of pixel-based data. These structures became the source of category definitions and served not just as an example but also as abstract normative placeholders for the class. In mapping visual objects to representational structures, this approach sought to reduce complexity by limiting the range of variation found within samples. In establishing a small set of parameters for what was permissible as possible variation, the number of features were dramatically reduced from all the possible pixels contained within an image or a region of interest to those contained within the set of structures and the distances between structures. This had the effect of redefining the visual space capable of representing objects from a field of symbolic values into a normative set of component parts organized around a placeholder category object. The creator of these objects then packages together his or her assumptions about this placeholder, the component parts, and the possible range of variations into the resulting model. This generates a normative field whose possible variations are regulated by the placeholder object.

The pictorial structures paper also produced a refinement of Firschein and Fischler's previously introduced primitives concept, as we can see in their account of the primitives that make up a typical image of a person:

> As an example, suppose we want to describe a frontal view of a standing person. This visual object could be decomposed into six primitive pieces: a head, two arms, a torso, and two legs. For this visual object to be present in an actual picture, it is required that these six primitives occur (or at least that some significant subset of them occurs), and also that they occur within a certain spatial relationship one to the other—that is, the legs should be next to each other, and below the torso; the torso should be between and below the tops of the two arms; and the head should be on top of the torso.[25]

Their normative assumptions of what makes up a person and the allowable variation within the category "person" are composed through the creation of this model. A hidden object, the abstracted placeholder human, sits at the center of this pictorial structure. Like a movable wooden reference manikin used in figure drawing, this abstract human occupies the place of a generic person. The model's creator builds his or her own sense of the average human into the pictorial structure. It would need to represent an average of the expected features of the category "human" because the measurements of variation, in order to reliably detect and identify an individual human among others, must be meaningful in terms of both the differences among the provided images and between the pictorial structure and the individual image. As Roopika Risam argues, "The production of a universalist notion of the 'human' relies on defaulting to the aesthetics of dominant cultures and languages."[26] No average human model can escape the influence of dominant understandings of the human and its expected variation. One of the major assumptions of such a model is that the variation would not be randomly distributed throughout the model but rather would fall into the space defined by the "certain spatial relationship" between component parts.

The twin object detection problems invoked by Fischler and Elschlager, faces and military sites, while initially seeming to share little in terms of features or goals, are both governed by the logic of a required set of features that might appear in different configurations. Take the eyes, nose, and mouth. These three facial landmarks, projected against the abstract face model, can form various different arrangements within the bounded space marked by the right and left edges of the face and the hairline. Fischler and Elschlager describe their approach as the assembly of component primitive parts that make up a whole: "Many, though by no means all, visual objects can be described by breaking down the object into a number of more 'primitive parts,' and by specifying an allowable range of spatial relations which these 'primitive parts' must satisfy for the object to be present."[27] The language of the picture primitives would come to be one of the primary ways of discussing the automatic recognition of complex images.

Collections of these key features would be taken together to provide generic or prototype objects for pattern matching. The allowable variations, located within some predefined scope, were made possible because the linkages between pictorial structure elements were conceptualized as elastic. This material metaphor for

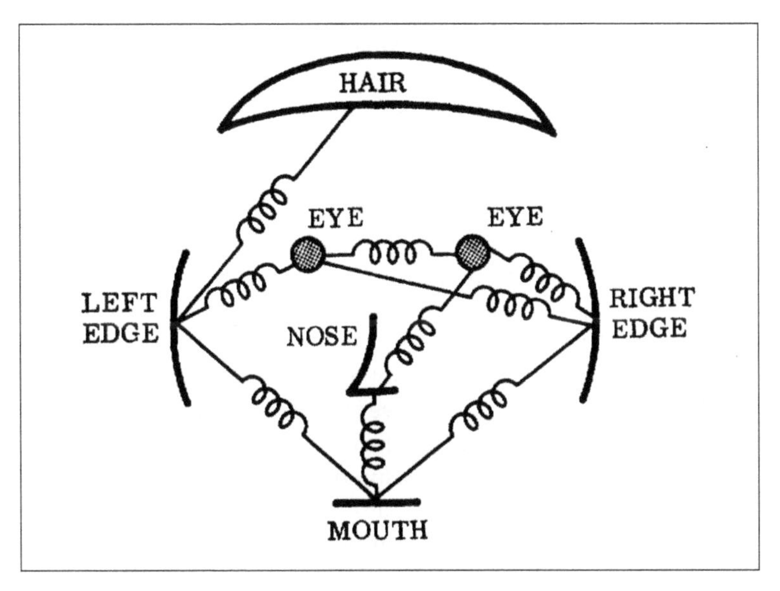

Figure 9. Pictorial structures for face recognition. From Martin A. Fischler and Robert A. Elschlager, "The Representation and Matching of Pictorial Structures," *IEEE Transactions on Computers* C-22, no. 1 (1973): 67–92, 77.

flexible relations among component parts defined their approach to this new form of pattern matching:

> Assume that the reference is an image on a transparent rubber sheet. We move this sheet over the sensed image and, at each possible placement, we pull or push on the rubber sheet to get the best possible alignment between the reference image on the sheet and the underlying sensed image. We evaluate each such embedding both by how good a correspondence we were able to obtain and by how much pushing and pulling we had to exert to obtain it.[28]

Their model for face detection defines an algorithm and metrics to measure individual differences as the distance from a referenced image. Although they explored the possibility of constructing a model that would contain all the variation found within the collected representations of human faces that would make up a face model, they decided instead to implement a model based on a flexible prototypical face. They did so by producing a refinement of their primitive concept. This prototypical face model, they claim, is an ideal model that would not match all faces but is instead something that would lead to a reasonable amount of matchiness, as found in variables measuring the difference between the singular and the prototype, with a range of faces. "We might consider all human faces," they write, "to be perturbed versions of some single ideal or reference face," but they could not address all the possible "noise and distortion" within such a model.[29] This notion of generalized objects sharing minor differences is common in the description of pattern matching and machine learning methods. The "ideal or reference face" is not an average face but something like a prototypical face. Normative facial features and the distance between these features provide the landmarks by which variations or "perturbed versions" are measured. In this account, all individual faces are perturbed versions, as no single face would match the ideal face template and each perturbed version of a face would have an associated transformation that would be required to transform or warp the space of an individual face to the universal face model. The metaphor of mechanical springs enabled them to develop a cost model, through the LEA (linear embedding) model algorithm, of the movement between component parts. A face would be described by a set of components linked together with springs: eyes, mouth, nose, left and right borders, and a hairline. Turning the face into what was essentially a topographical map enabled them to present complex objects as descriptions of feature sets and to locate and measure possible structures within the bounds of allowable "spring" between components. The greater the length of spring between components, the greater the cost for linking these primitive components. Finding similar costs for the connection of these primitives would produce lists of possible matches. Although an approximate or close match was considered good enough for the LEA model, the target of reasonable matchiness is not appropriate for any technology claiming to be capable of biometric identification and surveillance. Contemporary computer vision wields confidence scores as evidence of algorithmic correctness and effectiveness, but these numerical values often obscure the meaning of such matchiness and give the observer false confidence.

Fischler and Elschlager's paper makes the argument for the effectiveness of a topological approach to computer vision by demonstrating methods for both face and cartographic feature detection.

Like so many other computer vision projects, their method's targeted problem space primarily concerns the identification of objects in aerial surveillance and reconnaissance imagery. Fischler and Elschlager call these "terrain scenes," and the application of their method to Vietnam War–era military goals is made explicit in their

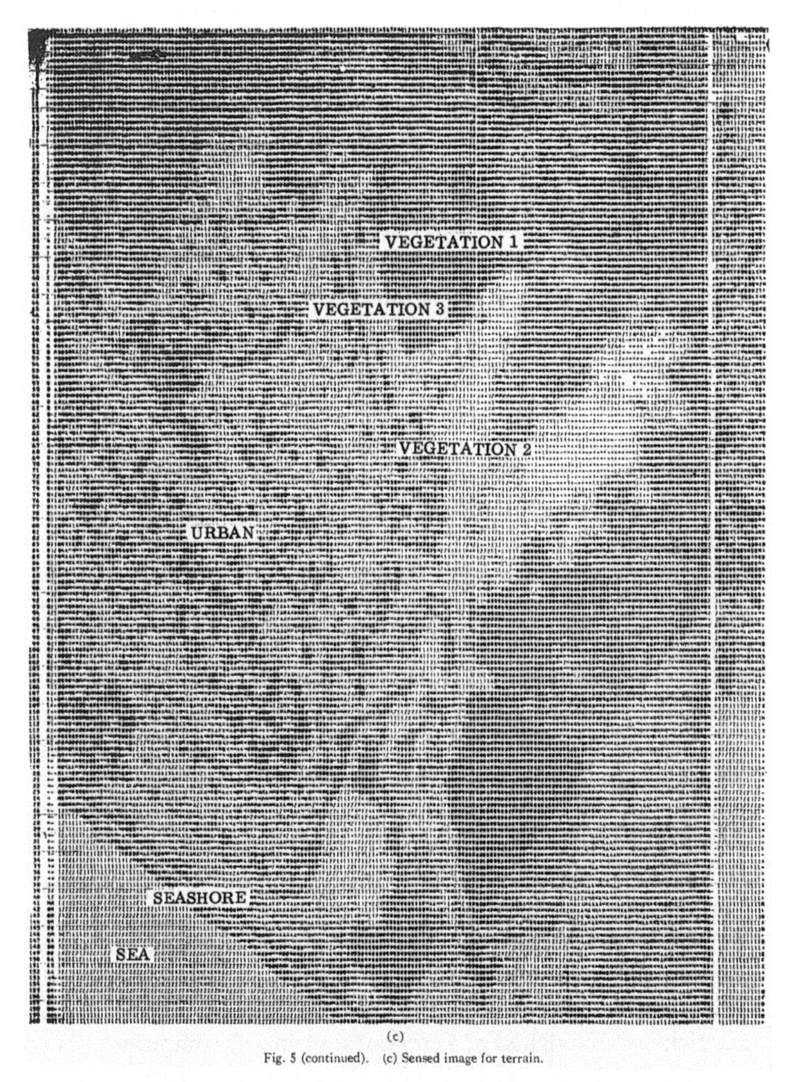

Figure 10. Pictorial structures for terrain recognition. From Martin A. Fischler and Robert A. Elschlager, "The Representation and Matching of Pictorial Structures," *IEEE Transactions on Computers* C-22, no. 1 (1973): 67–92, 90.

selection of sample scenes. Separating regions marked as vegetation, urban, seashore, and sea, they demonstrate possible solutions to locating and mapping geographical regions of interest from aerial photographs. There are many possible applications for this terrain model: finding forested areas for potential deforestation—methods that would essentially render Fischler and Elschlager's algorithm a virtual Agent Orange—and those transitional regions located near urban centers that might be harboring weapons and other military targets. Fischler and Elschlager's model of pictorial descriptions and pattern matching has become the foundation for many contemporary computer vision techniques, most crucially in the development of more advanced automatic face detection and matching systems.

Contemporary Methods of Face Detection

The most commonly used contemporary methods for identifying and extracting faces from images originate in the pictorial structures concept originated by Fischler and Elschlager. To better understand the history of this computer vision method, I will refer to the processes of generic face extraction as "face detection" and the matching of particular individual faces as "face recognition" or "face identification." If the pictorial structures method of face analysis can be said to depend on the creation of an average and normative face by which variations in terms of individual facial features are compared to determine the presence of a face (that is, face detection) or even individual faces (that is, face recognition), then some more contemporary methods, namely one known as eigenfaces, use a sequence of informative face patterns extracted from a set of sample faces to construct a similar normative set of faces. What computer vision researchers would call "face space" of both models is comparative in nature, which is to say that the overall numerical space of the model as well as the geometries of individual faces represented within this space are defined by the norms constructed from the selected faces. Matthew Turk and Alex Pentland, two researchers working in the vision and modeling group of the MIT Media Lab, describe their method in "Eigenfaces for Recognition," an article not published in a computer vision or computer science venue but in the *Journal of Cognitive Neuroscience*.[30] Turk and Pentland emphasize the practical applications of face recognition early in their article: "Computers that recognize faces could be applied to a wide variety of problems, including criminal identification, security systems, image and film processing, and human–computer interaction. For example, the ability to model a particular face and distinguish it from a large number of stored face models would make it possible to vastly improve criminal identification."[31] The eigenfaces method of face detection produces a set of representations of what they call standard pictures or eigenfaces that, when combined with a minimal amount of information from individual faces in the form of weights produced from the individual faces, can reconstruct and match identical faces. Turk and Pentland note that an eigenface resembles a "sort of ghostly face"[32] that can be brought into the sharper focus of an individual face by projecting the precalculated individual weights into this face space.

Another widely used contemporary algorithm for face detection, heavily indebted to Fischler and Elschlager's pictorial structures model, uses what are called Haar-like features to produce a regional bounding box, essentially a frame, around a set of pixels that have a high probability of being digital representations of faces. This algorithm is called Viola-Jones. It was named after its creators, Paul Viola and Michael Jones, the authors of a 2001 paper titled "Rapid Object Detection Using a Boosted Cascade of Simple Features."[33] This algorithm has been used to locate and focus on human faces in digital cameras, phones, and many other devices. The Viola-Jones algorithm searches for possible windows, or blocks of pixels, that might contain a face by matching patterns of segmented, smaller components already identified as faces. These smaller patterns are called Haar-like objects. Their method calculates the number of two, three, or four rectangular Haar-like patterns of pixels within a small re-representation of pixels that they call an integral image, then classifies these windows against identified face objects. While this method works with a variety of other objects and human part-objects that can be robustly classified using these patterns, its dominant use has been for face detection.

In a typical face-detection pipeline, a pretrained set of Haar-like cascade features is used to match possible pixels. These pretrained values have been evaluated on some data set of faces—faces not generally supplied with the trained values. This data set includes both positive and negative samples of faces. Positive samples would include a selection of possible faces, while the negative samples would be random, specifically nonface images. This construction of

Figure 11. Visualization of Haar-like features used for face detection.

a positive and negative example data set shows the persistence and prevalence of binary modes of thinking within computer vision. The OpenCV documentation explains that these cascade classifiers are ideally "trained with a few hundred sample views of a particular object (i.e., a face or a car), called positive examples, that are scaled to the same size (say, 20 × 20), and negative examples—arbitrary images of the same size."[34] The possible faces are thus defined in both positive and negative terms. The implementation of this algorithm in OpenCV comes with pretrained cascade data sets containing frontal faces, profile-view faces, faces with glasses, and even cat faces. These pretrained sets are tuned by the OpenCV developers to give better results for specific situations and body orientations, especially in uses when these orientations can be predicted. The construction of these data sets, like the faces selected for inclusion in an eigenfaces data set, defines what is a normative face; the ability of Haar-like feature face detection to find and extract a face depends on the similarity between the data set's normative conception of a face and the face found in a sample image.

This method of face detection mostly solves the preprocessing problem of "size and position normalization" identified by Marvin Minsky for a particular class of objects.[35] The bounding box surrounding the face enables extraction of the face from the pictured scene with enough standardization for comparison with other images. It also easily facilitates, by counting the number of these bounding boxes, an understanding of the number of faces and therefore humans present within an image. When combined with a sophisticated enough machine learning classification algorithm and a labeled set of faces, one can construct a tool for the identification of persons or for the classification of an individual face into subgroups, for example by race or gender. As imaging and surveillance technologies continue their spread from streets to borders and from our phones to our homes, concerns about the pervasiveness of face detection and recognition methods are increasing.

Figure 12. Picture of the author with a bounding box enclosing a detected face.

From Image Understanding to Computer Vision

The path from scene analysis and photointerpretation to computer vision as an organized field depended on the creation of an interdisciplinary area of research known as image understanding. This reconstructed conception of image understanding was understood to involve the grounding of extracted patterns, the filling in of missing components within and without the frame, that could only be supplied with subject-area expert knowledge. Image understanding named what was missing from scene analysis: a greater sense of the overall meaning of the images. The need for the fragmentation of the image into small component parts was realized during the phase of research known as scene analysis. This form of analysis operated mostly on discrete parts; the organized but unconnected relations between these parts remained incomplete until the development of new techniques for embedding this knowledge of images within algorithms and procedures. Once these components were in place, a reconceptualization of the field as computer vision would now be possible.

This shift in focus, however, also produced an ontological shift in computer vision. While the methods changed and images were increasingly fragmented into smaller units, the understanding of vision, perception, and the place of computation in relation to observation itself all underwent an ontological transformation. The image, although still treated as a representation of reality and objects within real space, was no longer connected to a possible human vision. As computer vision increasingly adopted the physiological model of human vision, it became less and less anchored to the simulation of a human observer. This change was contingent on a mobility of vision that brought computation increasingly closer to the sensed scene. This mobility involved the possibility of many arbitrary perspectives and image orientations. During this transformation, the image would become virtual and live, as it were, first and primarily in data structures where it could be instantly manipulated and transformed. Scene analysis thus names the transformation from a workflow dominated by image acquisition, digitization, and pixel-wise comparisons to one motivated by the long-wished-for possibility of sightless seeing. This goal of truly sightless seeing, however, would take some time to develop, as the automation of human perception was a tall order for the computing systems available in the 1970s—especially when this perception was understood to require a specialized body of knowledge derived from many previous observations and decisions.

Much like many other common sites of automation and computerization, the majority of early computer vision applications were created to replace the time-consuming manual labor of skilled and expensive human beings. The researchers, and more specifically their funders, were especially interested in automating tasks performed by expert military personnel. These projects were designed to reduce the amount of training required for a variety of technical occupations. The tasks performed by some of these military personnel would have been quite familiar to the developers of computer vision and artificial intelligence algorithms because the statisticians and computer scientists who produced the knowledge to found these fields were once directly employed by the military as photointerpreters. They were automating the tasks that were in essence the foundation of their field. It should come as no surprise that many of the basic statistical methods that make machine learning techniques possible were initially developed by photointerpreters and that these methods would later be used to modify the original tasks that inspired these methods. The historical entanglement of military photointerpretation, computer vision, and artificial intelligence would lead to revolutionary developments in imaging technologies for medicine and other fields that would improve countless lives, but the entanglement also contains traces of these earlier tasks while embedding logics that foreground classification, surveillance, and the targeting of difference.[36]

The extraction of faces, as we have seen, is an old area of research within computer vision. It is indebted to prior statistical analyses of the body made possible with film photography and what Allan Sekula calls the archival paradigm.[37] Computational face-detection methods were even produced from facial landmarks and biometrics before the creation of computer vision and the ability to directly analyze photographs.[38] The importance of finding and extracting faces and matching these with known photographed subjects was articulated early in the field's development. Faces were theorized as possessing a grammar, an organized logic that would be amenable to the dominant logics of segmenting a larger object into smaller component objects. These logics functioned under universalizing terms that posited a single model of the complicated object to be segmented. This model was limited in many ways. Its understanding of the human was based on a set of normalized topological features, or landmarks, that were taken as universal, and samples of these faces were generally provided from a small number of phenotypically similar people. The model was also limited in its understanding of embedded scenes; separate models had to be created for different photographed perspectives and configurations, with the front-facing image assumed as the standard. Classification systems, as Wendy Hui Kyong Chun argues, "require the prior construction or discovery of 'invariant' features, on the basis of which they assign and reduce objects."[39] These invariant features, in the above-described face models, are generated norms from collections of composite features that are then deployed as key informative features for future classifications—an especially insidious form of discrimination functioning as recognition.

The ARPA-funded Image Understanding Project was the site of much early computer vision research. The scope of this research and the ambitions of the researchers increased over time as the technology and methods improved, and as military needs intensified as a result of the ongoing activity in Vietnam. Image understanding became a broad area that combined expertise in traditional image-based activities, such as the military photointerpretation task and cartographic analysis, and newly developed techniques, including scene description. Over a twenty-year period, many projects were funded under the umbrella of image understanding. This area of research was fundamental to the development of the fields of computer vision and artificial intelligence.

The computer scientists working on these early pattern-recognition systems began their work with the explicit goal of replacing the human operators required for photointerpretation, or at least reducing their number. The move from photointerpretation to image understanding maps the transformation of early artificial intelligence systems onto what were called expert systems. Artificial intelligence found success only when it backed away from the promises of completely automated activities and was used as a decision-making resource for human operators. The move from the goals of complete replacement of photointerpreters with computer-aided object recognition was tied to technical difficulties as well as two social factors, namely resistance on the part of the interpreters themselves and the research climate of the 1970s and 1980s, in which artificial intelligence systems were increasingly replaced with knowledge databases and "expert" systems.

It was quickly acknowledged by the various project teams working on the photointerpretation problem that a focus just on images and object recognition would not come close to replacing the tasks accomplished by photointerpreters. Detailed and historical knowledge of the geographical area of concern as well as a larger organization of military movements and weaponry were required to make extracted visual features useful and to understand how they fit together. The ARPA Image Understanding Project was the continuation of research into automated object description and identification for photointerpretation started by Holmes and continued by Martin Fischler and his collaborators. This project was initially funded for the three-year period 1976 to 1979 with a grant of $1.3 million. Peter Hart served as the director of the artificial intelligence center at SRI International, the home of the program, and Harry G. Barrows was the initial principal investigator. The goals of the project were explicitly related to military intelligence and were defined in the abstract for the technical reports:

> The ARPA Image Understanding Project at SRI has the scientific goal of investigating and developing ways in which diverse sources of knowledge may be used to interpret images automatically. The research is focused on the specific problems entailed in interpreting aerial photographs for cartographic and intelligence purposes. A key concept is the use of a generalized digital map to guide the process of image interpretation.[40]

The integration of multiple sources of knowledge was as much a function of the complexity of the problem as a social solution to technological limitations of early computer vision. The wide net of intelligence gathering used to describe and categorize images has as its analog the same surveillance strategies used by covert operations and military reconnaissance forces. The form matches its function. In wanting to outsource the activities of photointerpreters to machine surrogates, these researchers needed to conduct close surveillance of surveillance experts; indeed, this is the logic of the recursive war machine. The task of photointerpretation, the classifying of aerial photographic images, was framed by project goals that were located in what would become a sprawling system. In a later annual report, Martin A. Fischler, who by 1978 had become the program's principal investigator, refines these project goals through the development of a system that would host the "generalized digital map" by linking together gathered knowledge of particular events: "In the present phase of our program, the primary focus is on developing a 'road expert,' whose purpose is to monitor and interpret road events in aerial imagery."[41] This road expert would be an automated photointerpreter, one with a constrained focus on vehicles and the roads on which they moved. Yet it was hoped that the ability to locate roads and to detect the movement of vehicles on these roads would be enough to replace these human experts.

The resulting automatized system was called Hawkeye. The Hawkeye project followed Fischler's previous insight that development in image understanding would require multiple data sources and that these data should be easily queried and made available for human decision-making. "We see," the researchers write in their 1978 annual report, "the military relevance of our work extending well beyond the specific road monitoring scenario presented above." The SRI road expert system, they argue, could be used for much more than just observing vehicular traffic. They suggest four other areas in which this system could be deployed by the military:

Figure 13. The Hawkeye system console. From Harry G. Barrow, "Interactive Aids for Cartography and Photo Interpretation," semiannual technical report, contract DAAG29-76-C-0057 (Menlo Park, Calif.: SRI International, 1977), 3. Image courtesy SRI International.

(1) Intelligence: monitoring roads for movement of military forces
(2) Weapon Guidance: use of roads as landmarks for "Map-Matching" systems
(3) Targeting: detection of vehicles for interdiction of road traffic
(4) Cartography: compilation and updating of maps with respect to roads and other linear features[42]

The SRI road expert and the Hawkeye system that were being sold to ARPA were well on their way to becoming generalized expert systems for military applications. This scene provided a prototype for today's computer-driven object tracking and remote control drone warfare system.

The researchers realized early in the development of their project that their expert system would need to focus on small tasks that would make the goal of photointerpretation possible for a human operator. The SRI researchers write, in their earlier 1977 semiannual technical report on the Hawkeye system to their ARPA funders, "We recognize that it is not possible to replace a skilled photointerpreter within the limitations of the current state of image understanding. It is possible, however, to facilitate his or her work greatly by providing a number of collaborative aids that alleviate the more mundane and tedious chores."[43] In a later report, Martin A. Fischler writes of the new tasks as follows:

- Finding roads in aerial imagery
- Distinguishing vehicles on roads from shadows, signposts, road markings, etc.,
- Comparing multiple images and symbolic information pertaining to the same road segment, and deciding whether significant changes have occurred.[44]

This trajectory mirrors the direction of the field of artificial intelligence as a whole: a move from the goal of totally automated analysis and interpretation to the production of expert systems that encoded a series of rules followed by human decision makers.

Expert systems emerged as a solution to the difficulties encountered in early computer vision and artificial intelligence in general. These systems encoded expert knowledge in the form of rules and coded data. Richard O. Duda and Edward Shortliffe's explanation of the shift to expert systems from artificial intelligence turns mostly on the difficulty of encoding, of quantifying, certain kinds of data rather than the impossibility of replacing trained human experts. For Duda and Shortliffe, certain cognitive tasks including "planning, problem-solving, and deduction" have been found to be hard to express in numerical form, and thus they want to reposition artificial intelligence as "consultants for decision-making" rather than the site of decision-making itself.[45] In order to make this shift, researchers needed to invent methods to standardize the presentation of information used as criteria for decision-making.

H. M. Collins, a sociologist and critic of artificial intelligence, understands one of the major moves of expert systems to be the formalization of acquired knowledge. He characterizes this as the move from the tacit to the explicit. The production of knowledgeable or thinking machines requires making explicit previously absorbed knowledge and experience. Once this tacit knowledge has become explicit, formalized rules for information processing can then be created.[46] These lists of rules required specialized forms for expression and evaluation and commonly were developed in the LISt Processor, or LISP programming language. The use of LISP and the focus on symbolic representations provide strong markers of the Hawkeye project's move into expert systems and the formalization of decision-making processes. LISP was created at MIT by John McCarthy, organizer of the Dartmouth Summer Research Project on Artificial Intelligence, and was the language of choice for encoding and processing information stored in lists.[47] SRI would later incorporate specialized hardware designed especially for executing LISP code—the same computing devices that others working in artificial intelligence rapidly adopted.

The military's desire to reduce the training required for technical staff was linked to wartime initiatives that required rapid mobilization and was especially intense during the Vietnam War. The cyclical dance through modularization, generalization, and specialization that characterizes developments in computer research and the computer industry matched similar social formations found in mid- to late twentieth-century institutions, including the U.S. military. Computer-driven automation is possible only through the combination of management techniques and technology. Although technology enables shifts in the techniques used and dramatically increases the speeds of these tasks, they are always already embedded within larger social systems, described and organized through the application of management strategies. When linked to the implementation of new technologies, these management techniques must involve some level of formalization of labor. Tasks are broken down into discrete components, changing the ordering and flow of labor required to perform the larger goal; this process frequently alters the outcomes of the original task. Automatization depends on the abstraction of the target task through managerial interventions and the introduction or layering of already existing alienated relations between the end product and the chain of humans involved in the design, production, and use of the automated tasks. In the case of one early and influential computer vision application, the activity of the person who was to be replaced was known as photointerpretation. Throughout the mid-twentieth century, the various branches of the United States military used these subject-area experts to categorize and interpret aerial photographs as part of their general surveillance and military planning activities. Theses photointerpreters carefully scanned the collected reconnaissance photographs to determine the contents and movement of both people and military equipment belonging to various adversaries.

Aerial photography, much of it performed before the creation of computer vision, was perfected for military use. The airborne gaze, from the bird's-eye view of a city or estate to photographs taken from airplanes, have long been connected with imperialism and the desire to dominate the landscape. These images were never objective, and aerial reconnaissance has never been an exact science. Historian of photography Beaumont Newhall writes of the task of the military photointerpreter, with special attention to the role of the imagination in reconstructing scenes:

> The techniques of photo interpretation are simple: the careful comparison of photo coverage over days, weeks and even months; the use of stereo vision; the measurements of images to the tenth part of a millimeter with a high-power magnifier fitted with a graticule; and, above all, visual imagination. The shadows cast by objects often reveal their profiles. Interpreters gradually became used to the view of the earth from directly above. To a good photo interpreter the identification of everything

> in the photograph became a challenge; "I don't look for things; I let the photographs speak to me," one of them once said. To search for unknown installations requires much ingenuity and imagination.[48]

From the creation of this occupation, military photointerpreters have facilitated the creative destruction of the war machine. Their expertise took time to achieve and required great familiarity with the landscape. At the same time, this knowledge was indispensable for demonstrating transcendent abstract power over a territory, especially when that territory was a remote foreign country. Collecting a stream of images and processing them quickly enables those using these images to locate changes that suggest possible movements of an adversary and potential advantages by anticipating actions from above. The military perspective of aerial photography also renders the territory a catastrophized geography. The view from above, as Caren Kaplan argues, is empowering and world making and yet at the same time "still a fragmentary thing, only barely holding together, despite the vigorous workout it receives in contemporary culture."[49] That fragmentation identified by Kaplan is registered in the split viewpoint of computer vision, the binocular machine and human vision that haunts the scene as a displacement of the knowledge of the photointerpreters. These phantom figures fragment the view from above. The computer vision–enabled aerial view brought new capabilities to those observing, but because of the rudimentary sense of structures used to analyze photographs, those under surveillance could deceive computer vision more easily even than the human photointerpreters. As Kaplan reminds us of contemporary drone imaging, "It is important to keep in mind that the high-tech enhancements of vision that this conglomeration of technologies and practices offer do not eliminate completely the 'fog of war' or the inherent tension between what a human 'sees' and what they 'know.' "[50]

Expert human photointerpreters were employed by the military to examine thousands of photographic images searching for known objects of interest. These objects were organized into a set number of patterns. These patterns were object matched or shared major characteristics with the rough patterns or types recalled by the interpreter. The shape of an airplane with its fuselage and wings, for example, would form a major "type" for manual pattern matching by photointerpreters. Collections of these types collectively made up the class of objects that represented different configurations of airplanes. Frequently the photointerpreters worked from grainy, low-resolution images taken from surveillance aircraft. Without the ability to determine the specific manufacturer and model of aircraft in a photograph, they would make use of the general characteristics of the object—basic shapes and outlines. Some of the most basic and foundational of contemporary computer vision techniques use the approaches and algorithms originally developed to solve these Vietnam War–era photointerpretation tasks. In their conception, operation, and description, these methods draw on the resources provided by the metaphors created during the original military framing of the task. The emphasis on image segmentation and mapping interobject relations within top-down aerial photography was required for the development of scene analysis and image understanding.

By the publication of their 1983 annual technical report, the SRI International researchers had largely replaced the language of image understanding with machine vision. The use of "artificial intelligence" to demarcate certain methods and algorithms was increasing, and these two research areas were now becoming explicitly linked. In this final report, the researchers wrote with confidence that they were nearing the completion of the project begun in 1976. They had constructed a test bed Hawkeye system at the U.S. Army Engineer Topographic Laboratories at Fort Belvoir in Fairfax County, Virginia. They had developed several new algorithms and explored the other possible military applications of their methods. They had embedded some of the latest programming languages and the most advanced computer hardware developed at the time in their test bed system. Most importantly, they had discovered some of the major technical and philosophical limitations of the emergent field of computer vision.

In September 7, 2010, a group of SRI International researchers including Martin A. Fischler were awarded patent US 7,792,333 B2 for a novel method of person identification.[51] Their patent proposed a method for matching and detecting individuals based on their hair patterns, as seen from overhead. This patent, which is admittedly a rather unusual proposal and seemingly easily defeated with a gust of wind or a hairbrush, combines Fischler's previous work on topographical models and top-down aerial imagery to match images of "hair boundaries on nonbald individuals by applying a line-texture operator, such as those used for extracting rural roads from satellite imagery, to the obtained image."[52] Like the face recognition methods above, this patent depends on some rather normative assumptions about hair, in particular the dependency on lines that indicate assumptions about the prevalence of relatively straight hair, which are perhaps not as widely shared as the authors of the patent imagine. The symbolic abstractions used in object, face, and even hair recognition are based on a series of cascading assumptions that are grounded in the assumption and presentation of normative objects and bodies. These norms are first imagined and then constructed by the methods and from the selection of "typical" cases. Computer vision, in its earliest moments and in the present, cannot escape from its reliance on symbolic abstractions and the biases, exclusions, and historicity that such modeling activity inevitably introduces.

« 4 »

Shaky Beginnings

Military Automatons and Line Detection

The mutual development of the fields of artificial intelligence and computer vision, as the previous chapters have demonstrated, was not a coincidence. In order to make progress on the complex and sophisticated tasks that were governing research in these two fields, the specialized techniques and strategies of both areas of research needed to be brought together. The combination of different sources of knowledge and perception was a key feature of this moment in computing and engineering. The Hawkeye system, SRI International's aerial surveillance research project, was imagined by its funders and designers as operating remotely—its name conjures an imaginary bird's-eye view of remote landscapes—yet this sense of remote control was only possible through the combination of multiple sources of information into a virtual photointerpretation workstation. This system necessitated a temporal and spatial shift of imaging data from the scene of surveillance to the virtual scene of computer-aided decision-making. This delayed processing of image data functioned within the general mode of midcentury computerization of intelligence gathering and information processing that itself was an extension of already existing military and civilian surveillance regimes. Hawkeye's real novelty was the centralization of gathered knowledge—knowledge acquired from both archived visual, which is to say scanned photographic, images and textual sources—and information into a database-driven and interactive workstation operated by domain experts. Artificial intelligence techniques were not yet advanced enough to recognize many objects of military interest and to automate many costly and critical decisions. This required decision-making to remain in the hands of human experts. Separating decision-making from the site of image capture introduced delays and made modeling the visual data of complex scenes much more difficult.

The goal of bringing computation directly into military action—in other words, placing the site of automatic, computerized decision-making directly within those spaces referred to by researchers and funders as hostile climates—remained elusive. It was difficult not just because of the impracticalities of midcentury computing with oversized hardware and slow image acquisition devices, but also because it involved a wholesale paradigm shift within visual sensing systems. The imagined visual space was no longer organized according to two-dimensional aerial photographs but a movable three-dimensional space that was itself the site of computation. This paradigm shift, which was already becoming active in the invention of elastic pictorial structures, altered the trajectory of computer vision and transformed the scene of computation. This paradigm shift was also an ontological transformation. It was required because the military needed photographic images to be rapidly processed. Improved algorithms and advances in the speed and capabilities of computing hardware helped make this transformation possible, but it was hardly a smooth transition from the temporally delayed analysis of digitized photographic images to the integrated site of computation and image acquisition.

The Shakey Project

Surveillance, mapping and navigation, and early embedded computer vision all converge in a mid-twentieth-century SRI research program called the Shakey project.[1] The Shakey project was officially sponsored by U.S. Department of Defense funding agencies for almost a decade, from 1966 until 1972. Work on some of the key technologies had begun before the awarding of the first contract, and the project would continue to shape the trajectories of computer vision and artificial intelligence for years after its formal conclusion. Culture and technology were in rapid flux during these years; the researchers and organizations involved would undergo changes in priorities and politics as campus-based protests against the Vietnam War intensified. When the Shakey project began, SRI was known as the Stanford Research Institute, and by the time it concluded, the institute was no longer affiliated with the university. This influential research project popularized robotics, introduced new planning and mapping algorithms, developed the ubiquitous line detection algorithm now used to keep cars between the lines on highways, and, like almost all the technologies discussed in this book, was designed for explicitly military goals. While the computer-aided knowledge workstation was the material product and the imaginative site for the Hawkeye project, the Shakey project was organized around the creation of a semiautonomous robotic device, affectionately called "Shakey the Robot" by Charles A. Rosen, one of the project leaders. Shakey enabled computer vision researchers to reconceptualize the task of pattern recognition and scene analysis by moving, at least at the theoretical level, the site of image recognition from a workstation to an embedded and movable device. The understanding of perception active in the Shakey project involved not just recognizing the content of images but also sensing the environment of perception—a key development in the history of computer vision.

A detailed and image-filled feature article written by Brad Darrach and published in *Life* magazine in November 1970 introduced Shakey to the broader American public. The article, which at times played on the same popular conceptions of a general artificial intelligence that had framed the announcement of Frank Rosenblatt's Perceptron, promised to give readers a sense of what Darrach called "the fascinating and fearsome reality of a machine with a mind of its own."[2] While Marvin Minsky, the MIT computer scientist and leader of Project MAC, had in previous years developed a reputation as a critic of machine learning—especially in relation to Rosenblatt's Perceptron—he is quoted by Darrach as giving what might be termed a rather optimistic account of the field's potential: "In from three to eight years we will have a machine with the general intelligence of an average human being. I mean a machine that will be able to read Shakespeare, grease a car, play office politics, tell a joke, have a fight. At that point the machine will begin to educate itself with fantastic speed. In a few months it will be at genius level and a few months after that its powers will be incalculable."[3] Minsky's predictions about the rapid rate of progress for artificial intelligence were questioned by the reporter, who was witness to some of the major limitations of Shakey, as well as some of the SRI International researchers quoted for the article. They were sure that these goals would be achievable in more like fifteen years.

A demonstration of Shakey slowly moving about the Menlo

Park offices of SRI International, navigating rooms and obstacles, provided Darrach with a dramatic scene that he used to take stock of the present state of computer vision. By embedding sensors in a movable robot and observing its movements, the SRI researchers staged for readers of *Life* a scene that was carefully designed and orchestrated to show machine perception in progress. The pauses and jerky movements of Shakey as it appeared to observe, plan, and navigate gave the impression of thought and volition. Perhaps one cannot help but anthropomorphize a robot that appears to pause while developing a plan that it then proceeds to put into action with slow, deliberate actions:

> Shaky *[sic]* was also thinking faster. He rotated his head slowly till his eye came to rest on a wide shallow ramp that was lying on the floor on the other side of the room. Whirring briskly, he crossed to the ramp, semicircled it and then pushed it straight across the floor till the high end of the ramp hit the platform. Rolling back a few feet, he cased the situation again and discovered that only one corner of the ramp was touching the platform. Rolling quickly to the far side of the ramp, he nudged it till the gap closed. Then he swung around, charged up the slope, located the block and gently pushed it off the platform.[4]

Perhaps it is only the use of "rolling" to signal movement, or the presence of this singular "eye" that makes us feel that something is slightly off as we read Darrach's account of Shakey's actions. His will to grant subjectivity and the presence of a mind to Shakey, to render this programmed "it" a thinking "he," is licensed by the optimistic yet unrealized prognostications given by Minsky and others.[5] This *Life* magazine profile of Shakey and the SRI researchers followed some previous minor coverage of the project in 1968 in the *New York Times*.[6] It seems likely that the *Life* article was perceived by SRI as an opportunity for positive public relations for the institute and to present Shakey as an innocuously potential office assistant rather than a war machine—the slightly different imagined application than had been originally articulated to Shakey's Department of Defense funders.

By the program's conclusion in 1972, Shakey had launched many other developments in robotics and in artificial intelligence, including a follow-up robotics project known as Flakey.[7] Shakey's importance to the field of computer vision, however, has yet to be fully recognized. This is because Shakey was primarily considered an artificial intelligence success, and in several accounts of the project, the computer vision aspects tend to be minimized. Shakey was designed to satisfy a specific set of conditions within a specific scene or environment. It was going to bring together the latest technology and innovations from several different fields in order to produce a computational representation of three-dimensional space. Shakey would be not just a single experimental device but also a platform for artificial intelligence research. The complex system of wired-together existing computers that would come to be known as Shakey the Robot was supported by two Department of Defense contracts. The first was awarded in 1966 as a result of a proposal written in January 1965 titled "Application of Intelligent Automata to Reconnaissance." The Stanford Research Institute was awarded contract AF 30 (602)-4147 by the Air Force for what was to become SRI Project 5953. This contract was administered by an organization quite familiar to SRI researchers: the Rome Air Development Center of the Griffiss Air Force Base in Rome, New York. The second contract was sponsored by the Advanced Research Projects Agency, or ARPA, and awarded to SRI in 1969. It was administered by NASA under contract NAS12-2221. This second phase of the Shakey project was called "Research and Applications—Artificial Intelligence," and it more modestly proposed to "investigate and develop techniques in artificial intelligence and apply them to the control of a mobile automaton, enabling it to carry out tasks, autonomously, in a realistic laboratory environment."[8] In 1971, in the final year of the second phase of the project, SRI researchers created an entire second Shakey robot system, implemented several new algorithms, and changed the back-end computer hardware from a Scientific Data Systems' SDS 940 to a Digital Equipment Corporation PDP-10.[9]

The new aspects of computer vision and the new orientation toward the image required for this project would enable an important shift in perspective, one that would produce the conditions of possibility for a platform—a test bed, as Peter Hart describes it in his "Making Shakey" contribution to the *AI Magazine* article "Shakey: From Conception to History":

> The proposal that launched the Shakey project was submitted by the Artificial Intelligence Center of Stanford Research Institute (now SRI International) in January, 1965. SRI proposed to develop "intelligence automata" for "reconnaissance applications." But the research motivation—and this was the inspiration of Charles A. Rosen, the driving force behind the proposal—was to develop an experimental test bed for integrating all the subfields of artificial intelligence as then understood. SRI wanted to integrate in one system representation and reasoning, planning, machine learning, computer vision, natural language understanding, even speech understanding, for the first time.[10]

In their initial proposal, the researchers expressed their desire to use this project, as they wrote of their lofty goals, "to conduct the necessary theoretical research and to develop the hardware and software to control a mobile automaton in its performance of non-trivial missions in a real environment."[11] The degree to which they believed they could develop a system capable of operating in a "real environment" might have been registered in the humorous name selected for the project. Shakey was an unstable project for several reasons. The device, as might be expected from its name, moved through space in a jerky manner. This was not because of delays in planning, although these also added to the instability of the device, but rather the state of robotics in the 1960s. Shakey was initially loosely connected by a radio transmitter to a large and immobile SDS 940 general-purpose digital computer that needed to be physically located in close proximity to the robot. If we take "autonomous" to mean a lack of close attachment to immobile equipment, then Shakey was in no way able to operate autonomously. The project itself made progress toward the development of computer vision in a haphazard fashion, and the results were frequently far less impressive than those promised in the project proposals. The Shakey robot and the larger project were designed to satisfy DARPA's requirements for "automatons capable of gathering, processing, and transmitting information in a hostile environment."[12]

Hart's evasive shift away from the desired military uses of Shakey to describing the project's motivation as basic research is typical of this group of researchers in recounting the history of the project and their involvement in this DARPA-funded research program.[13] Charles A. Rosen, Nils J. Nilsson, and Milton B. Adams were the authors of the initial proposal issued to their military funders. Rosen and colleagues organized their project around the following rather directed statement of purpose: "The long-range goal of this program will be to develop intelligent automata capable of gathering, processing, and transmitting information in a hostile environment. The time period involved is 1970–1980."[14] Their designation of the operating environment as a "hostile environment" was not ancillary or even a future possibility but at the center of the proposal; it was clear from the beginning of the project that SRI was proposing to build a machine designed for use in active military operations.

Shakey would require the ability to sense its environment through a camera that could take a picture of the space in which the device operated and locate walls, objects of interest, and obstructions. It would need to produce what in essence would be a continually updated map from a stream of digitized images of its environment. The project produced a change in the spatial orientation of computer vision by developing visual models of space based not on two-dimensional images but on a situated sensing device. In

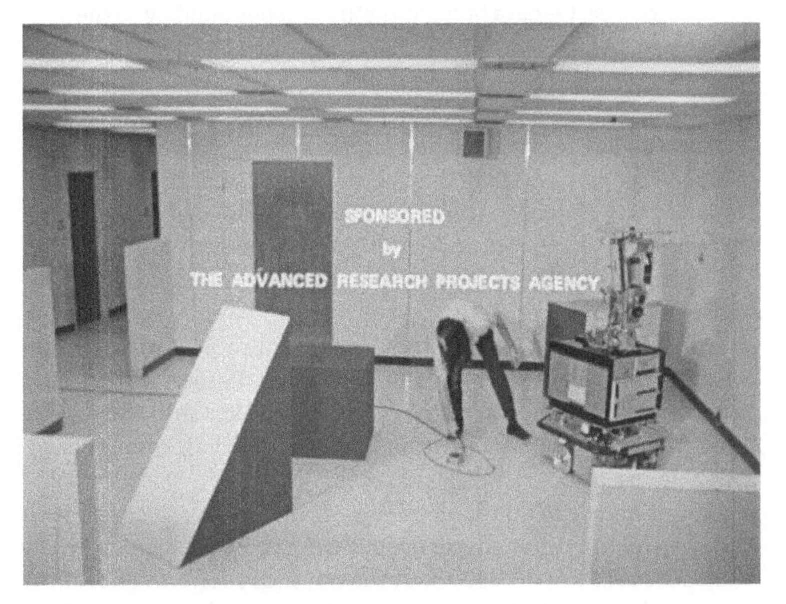

Figure 14. Title from Shakey promotional video (1972). The text overlay reads, "Sponsored by the Advanced Research Projects Agency."

the initial proposal, the researchers gave the term "internal model" to refer to the situatedness of the Shakey visual sensing device. This internal model of the space occupied by the sensing device, Shakey itself, would be constructed from multiple sensed visual scenes. The ability to reconcile these multiple images transformed the imaginary space of computer vision. This transformation is registered in the references to mechanical robots and automata that provided the background conceptual framing for the project:

> Mechanical robots that imitate human or animal behavior have intrigued man since antiquity. About 300 b.c., Hero of Alexandria constructed a hydraulically powered figure of Hercules and the dragon. The robot Hercules shot the dragon with bow and arrow, whereupon the dragon rose with a scream and then fell. Many robots embellished medieval clocks, such as the famous 15th century Venetian clock with its pair of bronze giants that strike the hours. These devices executed simple sequences of actions without change and, though entertaining, could hardly be called "intelligent." Twentieth-century automata—such as Grey Walter's turtles, Claude Shannon's maze-running rats and Johns Hopkins' beast—have brought the art of robot-building to the beginning of a new chapter—that of *intelligent automata*.[15]

In a retrospective account of the Shakey project, key researchers refer to Shakey less as a project and platform than an anthropomorphized robot. "Shakey the Robot" is the name they use several times, adding Shakey to their own blended genre mythic history of these mechanical robots and "intelligent automata."

Developments in computer vision were not an explicit outcome of the Shakey project. Peter Hart writes that work in computer vision was done out of necessity: "The initial project plan did not call for intensive research in computer vision. Rather, the plan was to integrate existing computer vision techniques into the experimental test bed. But, as it turned out, very little technology was available, so a focused effort in computer vision was started."[16] Shakey processed "digitized television picture[s]" from cameras installed on the robot. These images were processed and were the main source of knowledge about the physical environment for the system. "Line analysis," as Rosen, Nilsson, and Adams explain, "exploits the fact that the walls, doorways, and most of the other

Figure 15. Photograph of SRI researchers and Shakey in 1983. In back are Charles Rosen, Bertram Raphael, Dick Duda, Milt Adams, Gerald Gleason, Peter Hart, and Jim Baer; in front are Richard Fikes, Helen Wolf, and Ted Brain. Image courtesy SRI International.

objects in the automaton's environment have straight-line boundaries."[17] Being able to quickly detect and map the environment was essential for the mobile Shakey robot. As it moved around obstacles and through doors, Shakey needed to constantly update its stored map of the room. The early proposals did not outline specific algorithms or methods that would be used. The research into computer vision undertaken as part of the Shakey project was directed primarily toward the mapping of space through the movable yet still situated camera. This work would result in something more than a technical achievement through the development of new methods; it also involved an ontological shift that reconfigured the scene of digital image analysis. In moving the imaginative site of image analysis to a movable object, this project transformed the understanding of the image in computer vision. Image data were no longer conceived of as digitally acquired after the fact but processed in the moment. Without the physical presence of a human operator or photographer to aim the lens of a camera and establish a shot for later computer processing, Shakey's mobile frame was free to track and follow features detected in images.

Shakey and the Perceptron

It was initially envisioned that the computational activity supporting the visual object recognition components of the Shakey project would be powered, at least in part, by the SRI-developed local hardware implementation of Frank Rosenblatt's Perceptron algorithm. Stanford Research Institute researchers had created their own hardware implementation of the Perceptron, like CAL's Mark I, that they called MINOS, along with a later iteration known as MINOS II. Frank Rosenblatt was also hired as a consultant in the initial design of the MINOS machine.[18] While the MINOS II implemented Rosenblatt's Perceptron algorithm, SRI's learning machine was presented as an "evolved" device that incorporated insights from another competing machine learning project. That project was the ADALINE/MADALINE system, supervised by their Stanford University colleague Bernard Widrow. Added as an appendix to the initial proposal for the Shakey project was an early paper describing the MINOS II "learning machine." The SRI researchers initially wanted to use the already existing MINOS II, not just develop a single-purpose device, because they understood Shakey, which had yet to be named, as a system designed for integrating, as a platform, many of the previously disparate artificial intelligence research projects undertaken at SRI. Like the abstract Perceptron algorithm and its implementation as the Mark I, the MINOS II learning machine was developed with explicit computer vision goals in mind. The paper describing the MINOS II system, "A Large, Self-Contained Learning Machine," was delivered a conference in 1963 and coauthored by Alfred Brain, George Forsen, David Hall, and Charles Rosen, all of whom were employees at what was then known as the Stanford Research Institute.[19] While the creators of Shakey eventually decided to use a commercially available general-purpose computer, DEC's PDF-10, their impulse to bridge research in specialized machine learning hardware with computer perception in the form of a movable camera via robotics was realized in the software implementations of these algorithms.

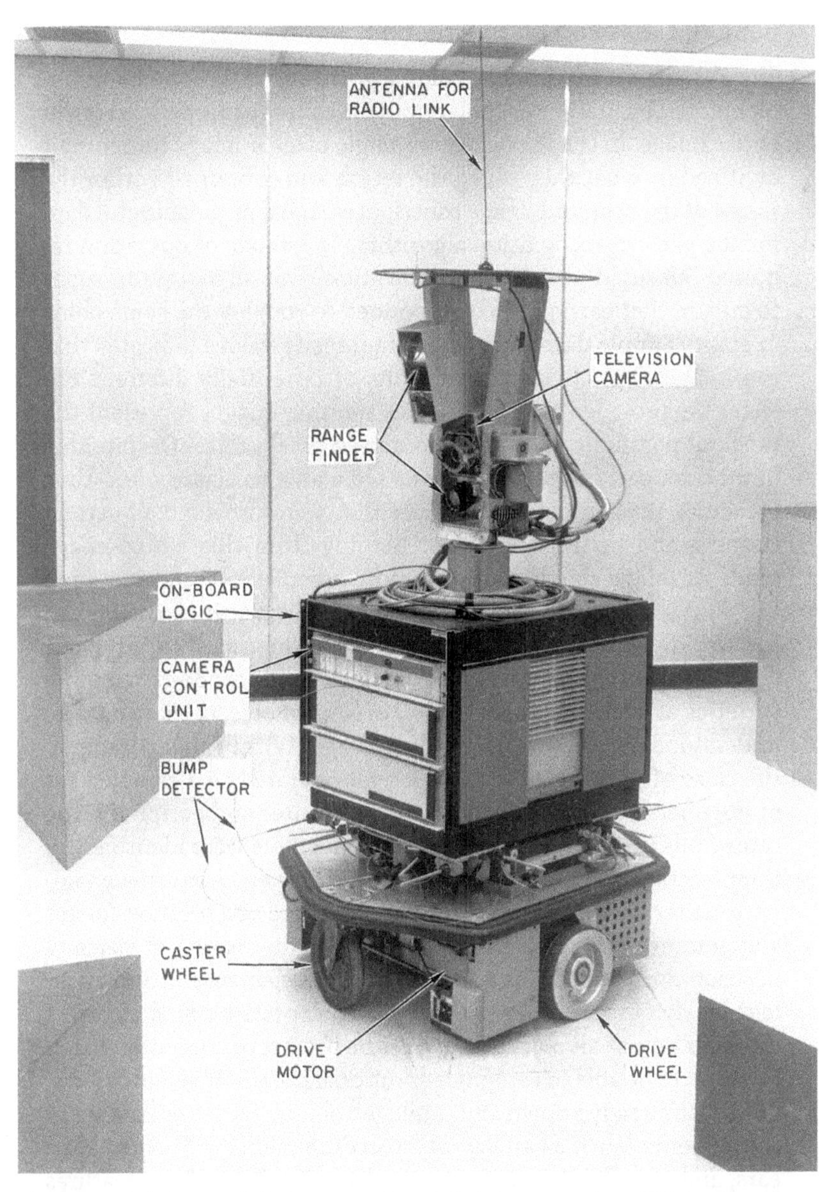

Figure 16. Labeled parts of Shakey. November 14, 1968. Image courtesy SRI International.

Feature Detection and Extraction

The Perceptron machine learning algorithm, as designed and implemented by Frank Rosenblatt and his colleagues, operated on an entire image. In this model, every single pixel or image component digitized by what they called the retina and contained within the frame of the acquired image contributes signal as meaningful data for the pattern-recognition algorithm. This mode of operation required "size and position normalization"[20] of all images in order to ensure that each pixel corresponded to roughly the same point in space. Simple difference in pixel intensity values indicates that the images contain representations of potentially different objects. The provided samples used as training images represent the range of possibilities of possible sensed pixel values. Despite this limitation, early implementations were able to classify and thus recognize images containing faces that were provided as part of training and testing data sets. Moving from this whole-image mode of image classification to more complex objects that were not all framed in the exact same way required several changes to the workflow. It also required rethinking the classification paradigm. This rethinking no longer operated in the mode of sampling varieties as such but the detection of known objects that could be understood as variants of an ideal image. Within this paradigm, the class of possible objects no longer operated as a sampled set of members but as a collection within a center and periphery. The center was imaginary, but all possible members were identified by their "distance," in terms of measured variation, from this imaginary center. Pictorial structures were invented as a solution for enabling more complex patterns and images that were not size and position normalized. The presence of these elements within an allowable distance from each other would contribute to the correct classification of an object. To move the project of computer vision forward, to be able to recognize additional classes of visual objects, not those already known and cataloged objects with a shared set of components (such as a face with its components of mouth, eyes, ears, and a hairline), but also more generalized features—features that could exist in different combinations within a much larger set of objects—was required. In order to detect this wider variety of objects, computer scientists needed to locate training data made up of a minimal set of features that could be reliably detected.

These core features, however, were all features that were shared by certain kinds of visual objects—objects that were primarily created by humans.

Digital images contain a lot of information. As the previous chapters have argued, with computer vision, these images are understood as symbolic representations of sensed scenes of reality. With pixel-based methods, it can be difficult to isolate particular sets of these representations. Even if images have been manually cropped or automatically segmented to provide a symbolic representation of a single object, the image data may contain information related to the background rather than the foreground object. Image data are collections of intensity values, and there is a great deal of variation among these values depending on the equipment used, lighting conditions, and variation within the field of vision. In order to begin any processing and analysis of images, some basic reduction of this information is required. Typical methods include the use of additional levels or layers of representation of the original image data. These representational layers provide access to reduced sets of image information to identify what might be the key components within the image. These sets of image information might not necessarily be the same values found within the image but some abstraction or transformation of those values. These reduced representation sets are commonly called "features." Feature extraction renders image data comparable. Conceptually, feature extraction techniques are computer vision methods that can be used both in preprocessing and in pattern-matching techniques. Computer vision methods should be understood as recursive; feature extraction might make use of machine learning methods to produce smaller sets of data that are then provided as input for the application of additional layers of machine learning to classify and recognize image content.

Two of the most fundamental and important methods of feature selection in computer vision are edge and line detection, two visual objects that are tied to human-created objects and useful for separating those objects from their surroundings. Edge detection has been primarily conceptualized as concerned with boundary detection, which is useful for the identification of more complex objects within images, and has been deployed in numerous applications, most recently as the drift detection technology used to alert drivers that their car has crossed the lines on the road. It has also been used as a preprocessing step in more complex applications as a method of data and dimension reduction.

The determination of what counts as a feature, even the selection of something as simple as basic geometrical shapes or lines, has major significance to the ontology of the image taken up by computer vision and to the realities of any application of computer vision techniques. In defining features, the developers and the users of algorithms and methods decide on the criteria that will be used to separate what they believe to represent meaningful signal from meaningless noise in image data. The search through image data for representations of lines should itself be considered a slicing operation. Cutting through image data to segment the meaningful from the meaningless draws boundary lines through data, partitioning and creating a new data representation. These boundary lines, not unlike the presence of straight lines themselves, reassert human-created divisions between what was taken as natural and what is designated as the product of humans. In collecting these extracted features and using them as input for higher-level machine learning, applications may inadvertently redefine the nature of the machine learning task through the exclusion or inclusion of potentially meaningful data.

The Hough Transform

Shakey's major contribution to computer vision was the improvement of existing line and edge detection algorithms, which are core algorithms used in historical and contemporary computer vision applications. Line and edge detection algorithms are required for basic pattern matching, image segmentation, and feature selection, as well as being a preprocessing strategy for more advanced techniques. It was recognized early in the development of the project that in order for Shakey to constantly recreate a map of the space in which it was situated, major objects and space boundaries would need to be quickly identified. Line detection algorithms were thus of special interest to the SRI researchers assigned to the vision components of the Shakey project, Richard O. Duda and Peter E. Hart. During their work on Shakey, Duda and Hart refined one particular line detection algorithm that has proved especially robust and important. Line detection algorithms were essential to the development of Shakey, and although Duda and Hart's paper on line and edge detection does not explicitly mention the Shakey project, their paper's major example problem—the identification of the presence of a box within an image—was absolutely central to their work on Shakey and the success of the project.

The Hough transform is now one of the most popular and widely cited methods for detecting the presence of lines in images and their combination as simple geometrical figures. This algorithm is classified by computer scientists as a transformation, or "transform," because it produces a plane transformation of the source image data in order to detect possible lines. The method that became the Hough transform algorithm was not initially intended to be used in digitized natural images; nor was it developed in the process of working within computer vision research, although possible applications of the technique were recognized by its progenitor. In 1960, Paul V. C. Hough, a physicist employed at Brookhaven National Laboratory, located in Upton, New York, on Long Island, was studying subatomic particle tracks from photographic images taken of bubble chambers at Brookhaven and the CERN laboratory in Geneva, Switzerland.[21] Bubble chambers were a recently developed instrument used in high-energy physics experimental research; they were invented by Donald Arthur Glaser in 1952, and he received the Nobel Prize for physics in 1960 for this device.[22] These encased chambers replaced the prior technology, which were known as cloud chambers, and were filled with superheated liquid nitrogen. An individual researcher could only examine a small number of pictures because as the number of these experiments increased, the work quickly became overwhelming. Hough submitted a patent for his method (U.S. Patent 3,069,654) on March 25, 1960, and it was awarded by the United States patent office on December 18, 1962.

In his patent application, Hough describes the significance of his innovative method of line detection and its application in particle physics research:

> This invention is particularly adaptable to the study of subatomic particle tracks passing through a viewing field. As the objects to be studied in modern physics become smaller, the

> problem of observing these objects becomes increasingly more complex. One of the more useful devices in observing charged particles is the bubble chamber wherein the charged particles create tracks along their path of travel composed of small bubbles approximately 0.01 inch apart, depending upon the specific ionization of the initiating particle. These tracks form complex patterns and are readily photographed with the use of a dark background. With this device, multitudinous photographs are produced with each photograph requiring several hours study by a trained observer to recognize the complex patterns of the tracks. It is therefore readily apparent, that as the photographs increase in number, the time consumed by a trained observer to study them becomes excessive and, unless large numbers of trained observers are used, the reduction of data falls far behind the production rate.
>
> It is one object of this invention to provide a method and means for the recognition of complex patterns in a picture.
>
> It is another object of this invention to provide an improved method and means for recognizing particle tracks in pictures obtained from a bubble chamber.[23]

The patent, as Hough's description makes clear, covers the use of this method for general pattern recognition in images, and in his ordering of the two provided objects, Hough gives priority to this application. At the same time, Hough's method was not tailored to the needs of natural images, and he details the specific scenes in laboratory research using bubble chambers in the remainder of the patent document. The track of bubbles in bubble chambers are the remainder of an event—an event produced by the movement of a particle through the superheated liquid nitrogen. The event itself can be detected by the presence of bubbles and momentum, "measured by the deviation of its track from a straight line in the chamber's magnetic field, and at high energy this deviation may be only a few bubble diameters."[24]

Hough's method altered the coordinate system of image data by manipulating data through a plane transformation. Line segments—thought of here as a series of points in the abstract geometrical sense, or as a series of bubbles in this particular case—appearing in the original image space will intersect in a point, or what Hough calls a knot, once they have been rendered through the plane transformation. This transformation operates on slope and intercept data. In Hough's work, line detection becomes a form of knot detection. Converting coordinates between the plane transformed space; the original image space enables one to expand a knot back into a line segment and thus recover the particle track. Hough's patent describes the use of several mechanical and computational devices to detect these particular tracks. Hough proposes that a television camera using an image orthicon tube would scan a photograph taken of the liquid in the bubble chamber and detect the presence of a bubble. This would produce a pulse that a custom electronic circuit would use as input for the above-described plane transformation. The result would be used as input for an oscilloscope, which would display the resulting knot. A second television camera pointed at the oscilloscope would detect, through a similar method as used by the other camera, the presence of a knot. The resulting coordinates would this time be recorded on magnetic tape for later transformation by a digital computer. This basic method of transforming the coordinate space of image data and locating intersecting points would soon be updated to simplify its operation and make it computable on more complex data.

Despite the primary focus on tracing bubble tracks in the patent, Paul V. C. Hough was not narrowly interested in physics. He delivered a general-audience talk as part of "A Computer Learns to See," a Brookhaven lecture series in February 1962 that introduced the basics of digital computers and described his method of detecting bubble tracks; he also provided some reflections on the relation between computers and human values. Hough's talk is remarkable for his combination of explaining the basic operation of digital computers with his speculative commentary on developments in pattern recognition, the generation of computer speech and art, and the role that computers may play in decision-making. The successful development of his technique for replacing the work of studying photographic images of bubble chambers gave rise to a set of questions about core human capabilities that appear to have troubled Hough. After explaining how information can be encoded as digital values and how computers might be used to generate new forms of art, he pauses to consider how what he presented as the generalized goal of computation—the maximization of a particular goal in terms of programmer-assigned incremental values to possible desired outcomes—should not be applied to some areas of life, even the aforementioned computer-generated aesthetic objects:

> At this point I tend to feel a value judgment coming on. I tend to say music and painting are for individual human enjoyment. Let the computer produce any set of paintings it wishes, and I will preserve one if I like it and even hang it up on some wall. Let someone even ingeniously program some painting rules into the computer and, again, if anyone likes the result, let him keep it. Best of all, for someone who likes to paint but simply can't draw a straight line, let him use the computer to make himself some paintings to hang on his own wall. But do not let anybody ascribe a value number to a painting, human- or machine made, and try to maximize that number.[25]

Hough invokes the emergent field of cybernetics through the notion of self-organizing systems as a form of learning, which he understands as the "highest human faculties" and thus linked to deep questions of value.[26] His concern with value questions as they apply to human- and machine-generated work is prescient, especially when one considers the capabilities of computing in the early 1960s and the rather simple forms of "seeing" enabled by his contribution to computer vision. His expectation, widely shared at the time, is that there will soon be rapid development of these machine learning technologies, and questions of value connected to human experiences of enjoyment, creation, and pleasure will become pressing matters. It does not take much imagination to think of Hough's invocation of a pair of television cameras in the patent application as providing him with a model of "binocular" computer-aided vision as an analog to human vision. While this delivers what we might more usefully think of as second sight rather than the physiological model based on simultaneous overlapping fields of vision, Hough's method of acquiring visual data makes possible seeing at what he terms "the threshold of computer capability."[27]

Hough's line identification and extraction method was extended, turned into an algorithm for digitized natural images, and popularized for widespread use in computer vision by SRI International researchers Richard O. Duda and Peter E. Hart. Their highly cited and influential article, "Use of the Hough Transformation to Detect Lines and Curves in Pictures," was published in 1972 in *Communications of the ACM,* one of the leading academic journals in computer science.[28] Duda and Hart's paper appeared at the end of the formally funded period of the Shakey project. In his retrospective account of the Hough transform published in 2009, coauthor Peter E. Hart explains the origins of this method of feature extraction. Hart recounts how Paul Hough's earlier work on the problem of line detection was initially brought to their attention through a brief aside in a state-of-the-field book on computer vision that had been recently published by Azriel Rosenfeld.[29] Rosenfeld, an expert on early computer vision methods and author of several field surveys, had mentioned Hough's work and this algorithm's possible application in computer vision as a method of feature detection. Duda and Hart's effort to modify the Hough transformation and extend the line detection algorithm to curves was supported by a contract from ARPA and NASA under contract NAS12-2221. This contract—it was given the more explanatory title "Research and Applications: Artificial Intelligence"—was the major funding mechanism supporting artificial intelligence programs at the Stanford Research Institute. This funding was connected to the Shakey program, and the Hough transform was essential for Shakey's ability to map and navigate through rooms and around objects.

Duda and Hart exchanged angle–radius values for the slope–intercept data used by Hough to limit the unbounded nature of the prior approach. As Duda and Hart demonstrated, using these values would make many more uses of the algorithm possible by reducing the number of calculations required to detect intersections in the transformed space. Duda and Hart introduced some changes into the terminology used by the transform, now calling the transform plane a more generic "parameter plane" and referring to the initial coordinate system as the "picture plane." This was done in part to fit the discursive construction of the algorithm to the needs of a generalized computer vision instead of those found in experimental physics. Rather than locating intersecting points or knots from lines in the parameter plane, when using angle–radius values, their revised approach would find the intersection of curves. They also implemented the mechanism suggested by Rosenfeld by which accumulated values in an array would be used to locate the knots or intersecting crossings.[30] Their paper introducing this updated Hough transform to computer scientists now assumes directly digitized image data as input and a completely digital environment; no longer would multiple scans of image data, oscilloscopes, and specialized hardware be required. With the entire workflow of the Hough transform running on a digital computer, Duda and Hart demonstrate several preprocessing steps that greatly enhance the ability of the algorithm to detect the presence of lines and curves. They provide as an example of this new workflow a series of images that show a still from a camera, a lower-resolution digitized image 120 × 120 pixels in size, a gradient image that has been filtered and reduced to binary values to mark the presence of possible edges, and finally an image of the lines detected through the process of transforming the gradient image into parameter space. Duda and Hart's method and their image preprocessing workflow has remained basically intact in the many years since its 1972 publication. The Hough transform can be found embedded in many devices, from digital cameras to automobiles. It is widely available in many popular computer vision packages used by software developers and hobbyists alike.

Contemporary open-source computer vision packages, including OpenCV, contain a basic implementation of Duda and Hart's updated Hough transform for use in line detection applications or as a method of feature detection. OpenCV includes the transform as the function HoughLines and another function, specialized for the detection of curves or circles, HoughCircles. These two functions operate on binary, single-channel images, meaning that source images, much like those used in Duda and Hart's demonstration, should be preprocessed to reduce color or gray-scale information in the image data to simple binary values. In many typical applications, the lines are detected from this reduced set of data in the binary image—it is much less computationally expensive to search a space of binary values, and the information reduction process produces distinctions at the edges of objects within the image—and then overlaid on the original full-color or gray-scale image rather than the binary image. Additional parameters added to these two OpenCV functions can apply a multiscale Hough transform to images.

The extracted Hough features can be combined to build training data sets to describe more complex objects. Training a machine learning classifier on the extracted configuration of lines from a larger number of samples of complex objects can produce reliable object detection systems. Using these trained classifiers, it is possible to tell, for example, if an image contains a relatively common and coherent object like a building. It can also identify common patterns in aerial photographs. The reduction of images to component parts that form comparable sets of features has been the primary method by which computer vision has addressed images since the ontological shift established by the research projects of the late 1960s. The mobile camera, as seen in drone images, has been the standard perspective of computer vision for decades. The top-down perspective of aerial photography, as was discussed in chapter 2, shaped the requirements for computer vision tasks. The detection of edges and lines was especially important to the understanding of these images, especially to those surveillance images that represented human-made objects. Because of the complexity of moving computers into the air in the 1960s and 1970s, the mobile frame was mostly land bound. In contemporary drone devices, even in low-cost, commercially available hardware, algorithms have migrated into the platform itself, continuing the effort begun with the Shakey project.

The best-known and most common contemporary use of the Hough transform would be lane departure warning systems (LDWS).

Figure 17. Line detection using OpenCV's implementation of the Hough transformation. Original image by Joseph Mehling, "aerial view," 2006. Dartmouth Digital Library Program.

These systems are typically manufacturer-embedded devices with sensors and processing elements that enhance vehicle safety. The Hough transform algorithm is the major algorithm used in LDWS systems. When a car begins to cross a detected line on the highway, usually in the form of lane markers painted onto the surface of the road, a visual or kinetic warning is given to the driver. The primary task of a LDWS is updating a model of the detected lines and predicting with a high degree of accuracy if the vehicle will soon cross one of these lines. These embedded systems that apply the Hough transform to a stream of images from a movable device render many contemporary automobiles the direct descendants of the Shakey project. Route-planning software packaged within GPS devices or supplied by smartphones are also indebted to the planning software developed for Shakey.

Falling Out with ARPA and Protesting Militarized Computer Vision

During the 1970s, a number of individual researchers, projects, and institutions involved in computer vision research began to change how they interacted with the core problems that until this point had defined most of their research. Some researchers came to applied research concerning computer vision from work on technologies that were military in nature; others had migrated their research interests toward the military because these were the only agencies and organizations with the money to purchase expensive computer systems and build custom hardware.

Nils Nilsson's career trajectory illustrates some of the ways in which institutional affiliations were altered and attitudes were changed by the war in Vietnam and the political exigencies of United States Department of Defense–supported funding agencies during this time. Nilsson, one of the major figures involved in the Shakey project and a coauthor of the proposal, had been on active duty in the Air Force and stationed at the Rome Air Development Center at the Griffiss Air Force Base in New York, the center that managed the initial contract for the Shakey project, before coming to California to work at SRI International.[31] Like other early computer vision researchers, Nilsson held an academic appointment alongside his employment at a commercial or military laboratory. Yet it was not the case that Nilsson unquestionably supported the growing demand from SRI's major funders that all projects should be focused on military goals rather than basic research. Shakey was pitched as a platform for embedded military surveillance, and that enabled the researchers to work on a variety of problems connected to the platform itself. As ARPA required increasing military applicability of their funded projects during the 1970s, some researchers sought to distance themselves from these goals. Nilsson describes how in 1974 he sought to pull away from work that had explicitly military uses by stepping away from APRA-funded projects: "We were also encouraged to have the system give advice about a more obviously military piece of equipment than the air compressor we had been using as a focus for our work. ARPA was simultaneously becoming more insistent about applying AI techniques to so-called 'command and control' problems."[32]

The insistence on direct applicability to military work from ARPA was the result of what is now known as the Mansfield amendment, two amendments attached to funding mechanisms sponsored by Senator Mike Mansfield of Montana. The first was applied to section 203 of the Defense Procurement Authorization Act, Public Law 91-121, and passed in 1969. This amendment directed the Department of Defense to halt their support of general or basic research and to only fund research directed toward military goals: "None of the funds authorized to be appropriated by this Act may be used to carry out any research project or study unless such a project or study has a direct and apparent relationship to a specific military function or operation."[33] The second amendment carrying this name was passed in 1973 and applied this same criteria to ARPA-funded research projects.

It was during this same time, during the Vietnam War, that students on several campuses—most crucial for this area of research on the campuses of Cornell University, MIT, and Stanford University—began to understand that the projects undertaken by institutional researchers at their universities were directly connected to military goals and the ongoing war in Vietnam.[34] The students, like some of the researchers themselves, began to question the now explicitly military applications of the new technologies and devices being developed within university and affiliated laboratories. While researchers working on computer vision generally wanted more freedom for their research and the ability to decouple military aims from the development of artificial intelligence and computer vision, the students were protesting the existence of this research at all on the grounds of their campuses. The opposition to research with explicit military goals was similar to the movement to remove ROTC programs from university campuses in that both activities involved what the students argued was an unethical use of the university. During this late 1960s and early 1970s moment, undergraduate and graduate students, postdoctoral students, and research fellows were routinely assigned to research laboratories that were housed within research institutions and required that they work closely with Defense Department officials. Support for much of the computer vision research undertaken by students and others was paid through ARPA/DARPA contracts and subcontracts from other organizations, including the United States Air Force.

SRI International, the present name of the research institute, was founded on November 6, 1946, as Stanford Research Institute by the trustees of Stanford University in partnership with several California corporations. The postwar economic environment made possible what had been previously a long-standing desire on the part of Stanford administrators. The institute was organized around an imperative to seek funding for scientific and technological research from the business community. Weldon Gibson, a longtime SRI executive and author of the SRI-published official multivolume history of SRI, writes, "We were expected to serve business and industry—first and foremost. Whether right or wrong in whatever degree, service to government was secondary. This may have been an all-too-simplistic view but our directors had no doubts on the matter."[35] The pursuit of governmental funding for military research, which would later dominate the activities of the institute, was not the founding intention. This corporate relationship would distinguish Stanford Research Institute from other university-affiliated laboratories of the period, including MIT's MITRE Corporation and CAL. Over time, however, SRI began to seek out military funding to support computer research that often required building custom hardware and expensive off-the-shelf computer systems from major vendors, including IBM.

Much like CAL in Buffalo, New York, was eventually split from Cornell University and established as an independent research institute, the link between Stanford University and Stanford Research Institute was officially severed in 1970, a tumultuous year of antiwar protest and activism on campuses across the United

States. Stuart W. Leslie terms this period the "days of reckoning," a moment of intensified political activity directed now inward toward the institutions themselves.[36] This was a moment when students and many faculty members organized together to demand divestment from military-funded research and the closure of university-owned laboratories that conducted research for the war in Vietnam. Stanford students had been opposed to the presence of SRI on the university campus for several years by this point. In April 1967, *Resistance*, a student newspaper, detailed in a cover story titled "Stanford Research Goes to War" the numerous links between the laboratory and the increasing military action in Vietnam and elsewhere.[37] From readings of publicly available reports, the article exposed Department of Defense funding for projects that included research into chemical warfare, guided missiles, jungle communication systems, and a series of contracts since 1966 for the development of aerial surveillance and reconnaissance systems that totaled almost $3 million of financial support. The students criticized the university for its fostering of the military-industrial complex through its development of Industrial Park, which housed several private military contractors:

> We get some insight into SRI's role in Provost Emeritus Frederick Terman's "community of technical scholars" (including the University, and the Industrial Park) when we note (as in our last issue) that corporations heavily involved in manufacturing Hawk and Nike-X, reconnaissance and surveillance systems, and airborne and ground-based military communication systems in Vietnam—such things as have been researched by SRI—are major tenants of the Park and are controlled by Trustees, SRI Directors, and members of the Stanford faculty.[38]

There were, as the article notes, many connections between these organizations, and there was little university oversight into the research performed at SRI, especially over those projects that were determined to be classified. It would take three more years after the publication of this article for just some of these links to be dissolved and for SRI to be formally separated from Stanford University.

There was growing awareness at Stanford that SRI's research had become far less basic and more highly focused and applied to goals outlined by the Department of Defense funding agencies. While Shakey was conceived in response to the desire for a military robot to be used in warfare, it was ultimately unsuited to such environments and reimagined for broader applications. There were many other SRI research projects, especially those not dependent on so many new mechanical components and sophisticated computational aspects, that resulted in knowledge and tools that were able to be applied to U.S. military operations in Vietnam. The institute reported that in 1968, $6.2 million of its $65 million budget was spent on projects in Southeast Asia.[39] SRI was known to be involved in developing biological chemical weapons and in "counterguerrilla projects," and concern with these projects was discussed in several venues.[40] After growing frustrated with the university's response to demands to divest itself of SRI, on Friday, May 16, 1969, hundreds of Stanford students went to SRI's Hanover Street office in the Stanford Industrial Park of Menlo Park to protest. The students blocked traffic at the intersection of Hanover Street and Page Mill Street and defaced the SRI building, painting "Research Stops Today" on a window.[41] One hundred fifty police officers came to the scene and teargassed the protesting students. Sixteen protesters were arrested.[42] In January 1970, Stanford's board of trustees approved a plan to sell SRI to itself.[43]

After the split from the university, the research laboratory was renamed SRI International. Despite this change in status and operations, SRI remained physically and intellectually close to the activities of the university. Students and some faculty were concerned that despite the now clearly established boundary between the university and SRI, these two institutions were still sharing resources. Richard W. Lyman, who was provost of Stanford University from 1967 until 1970 and then served as the university's president from 1970 to 1980, does not deny that SRI was deeply engaged in military research during this period, but he does describe the relations between the two institutions as quite loose:

> True, SRI was not created to do war-related research, though that eventually came to dominate their agenda. From a university standpoint, by the late 1960s the ties with SRI had become gossamer-thin. There were Stanford trustees who also served on the board of SRI—that was about it. A few faculty had individual ties to SRI. But the university's academic administration had

> nothing to do with planning or administering SRI's program—let alone paying for it.[44]

The students disagreed, and they had evidence. The anger directed toward the complicity of Stanford University in the military-industrial complex, of which SRI serves as a prime example, again boiled over when it was reported that SRI researchers were using university-owned computers to debug and run a war games simulation program called GAMUT-H. The existence of this program, and its authorship by an SRI researcher, was announced to the entire campus community in a guest article authored by Lew Lingg titled "SRI War Games at Computation Center" and published in the *Stanford Daily* on Tuesday, February 9, 1971.[45] The author of the article claimed to have had access to the source code listings for GAMUT-H—listings that included contact information for SRI and the name of the programmer, Andrew Grant. The article describes the function of the code and the nature of the military simulations and ends by issuing several demands to the university:

> SRI's use of the Stanford Computation Center for debugging their war game is another example of the University's covert complicity with the military-industrial complex. The Inquisition asks the Stanford Community to join in the following demands: 1) the University disclose all instances of utilization of the Computation Center and other Stanford facilities by individuals or institutions that are not constituents of the University, and 2) the University immediately terminate all provisions for such use on defense contracts.

The publication of this *Stanford Daily* column now provided a new target for antiwar protestors: the Computation Center.

On the same day the "War Games" article was published, a large group of students gathered for an evening discussion forum in which they debated the best way to respond to this news. The student debate was broadcast live on KZSU, the campus radio station.[46] One of the early speakers identified himself as working for the Computation Center and told the audience that the SRI program was one of the larger programs run on the Stanford computers and that it used up one third of the resources of this computer when it was running. After some discussion, a proposal was passed through a show of hands to meet the next day at noon and walk collectively over to the Computation Center. An estimated one hundred to two hundred students gathered at the Computation Center the next day, Wednesday, February 10, 1971.[47] H. Bruce Franklin, a tenured member of Stanford's English department, was one of speakers. Franklin told the crowd gathered in front of

***** Special Issue *****

Stanford Research Goes to War

RESISTANCE

Volume I, Number 2 — April 4, 1967

By David Ransom

SRI SAYS THAT IT WANTS TO WIN THE WAR IN VIETNAM. BOTH THE WAR AND SRI'S POSITION ON IT ARE PREDICTABLE GIVEN SRI'S HISTORY OF SIMPLE ANTI-COMMUNISM AND ITS ENCOURAGEMENT OF AGGRESSIVE FOREIGN INVESTMENT. IN FACT SRI PUBLICATIONS PREDICT THE WAR AS EARLY AS 1957 AND SUGGEST HOW IT WILL BE FOUGHT AND HOW IT OUGHT TO BE RESOLVED. SRI RESEARCH HAS TAKEN ADVANTAGE OF THE WAR NEEDS TO DEVELOP NEW AND IMPROVED WEAPONS AND WEAPONS SYSTEMS WHICH TENANTS OF THE STANFORD INDUSTRIAL PARK, CONTROLLED BY STANFORD TRUSTEES, DIRECTORS OF SRI, AND STANFORD FACULTY MEMBERS, NOW MANUFACTURE. STANFORD AND SRI HAVE HELPED DEVELOP THE METHODS AND THE MUNITIONS FOR CHEMICAL AND BIOLOGICAL WARFARE, INCLUDING THOSE USED IN VIETNAM. SRI MEMBERS AND SRI PROJECTS HAVE CONTRIBUTED TO MAJOR STRATEGIC DECISIONS IN VIETNAM, AND THEY ARE CONTINUING TO DO SO. THESE FACTS ARE DEVELOPED AND EXPLAINED IN THE FOLLOWING ARTICLE.

"To promote and foster the application of science in the development of commerce, trade, and industry...[for] the improvement of the general standard of living and the peace and prosperity of mankind."

—Stanford Research Institute

Both SRI and Stanford University deny they are intimately connected, but the truth is otherwise. In 1964, when the Government accused SRI of having swindled it of $250,000 (for charging the Government, as client, for depreciation on buildings which it had donated to the University, and which the University had then donated to SRI), the General Accounting Office remarked that the Institute had itself acknowledged "that the close ties with Stanford University are clear cut and unmistakable as evidenced by the fact that the trustees of the University are the general members of the Institute and elect the Institute's Board of Directors." The GAO also noted that the number of Stanford trustees serving as members of the board of SRI had increased from six to eight in the previous eight years, and it concluded that "the trustees of Stanford University, acting as general members of the Institute and as electors of the directors of the Institute, are in a position to exercise control of the Institute as well as of the University regardless of the Institute's statement to the contrary." Ernest Arbuckle, Dean of the Business School and a former Stanford Trustee, is presently Chairman of the Board at SRI, replacing Wallace Sterling, who remains a director.

SRI executives like to boast of their connections with the University. In a speech given at a banquet of the Bay Area Engineering Societies in 1959, Jesse Hobson, then President of SRI, said that "we participate in an honors graduate study program at Stanford University, contribute to the University's general funds, and provide assistance to

(Cont. p.2, col.3)

Corporations and The Men From SRI

By John Saari

FOLLOWING ARE PROFILES OF SOME OF THE CORPORATIONS WHOSE DIRECTORS ARE ALSO DIRECTORS OF SRI.

FMC

FMC Corp., San Jose, which at times has had as many as three Stanford Trustees and three SRI Directors on its board (presently: William Hewlett, Stanford, Paul L. Davies, SRI), is a major arms contributor to Vietnam and it expects its "defense" business to grow: "Development programs now underway give promise to provide important additions to volume in the years ahead and to further enhance FMC's position as one of the most experienced developers and producers of specialized defense material" (Annual Report, 1966).

PR men for FMC tell you "there isn't

(Cont. p.5, col.1)

Trust the Trustees?

By Robert Hass

The profiles which follow demonstrate some of the connections between the men who run Stanford University and the war in Vietnam. For those of us who are opposed to the war for moral and political reasons and who do not believe that the cause of the war is merely a foreign policy too rigidly anti-communist, the profiles may also be seen as the beginning of an effort to understand the causes of American foreign policy.

I do not mean to make the crude suggestion that the Vietnam War is only a creation of the American profit system, but two generalizations on the subject can be made:

(1) An American victory in Vietnam will have the effect of protecting the massive American investments in the underdeveloped

(Cont. p.6, col.1)

Figure 18. Cover page of *Resistance* 1, no. 2 (April 4, 1967).

the Computation Center that this site was a "good target."[48] The students occupied the site for three hours. Occupying the center was a deliberate strategy. It was what they called a strike and was intended to disrupt activities at the center and across the campus. It was designed to draw attention to the ongoing violence in Southeast Asia and the university's complicity. In his account of 1970s-era critical reaction to the technological, Matt Tierney argues that the computers found in Stanford's Computation Center provided protesters with local access to the remote war machines:

> To fill the computer center with the living bodies of scholars, from undergraduate to faculty, was to perform an inverted metaphor for the accumulated deaths in Southeast Asia. It was to produce a small amount of inconvenience for a great deal of noise. It was to name Stanford's research apparatus, its computers, as "the most obvious machinery of war." And it was, even without smashing, to silence that machinery.[49]

During the protest, the son of a faculty member was wounded by a gunshot. After the protest, the students' attention and protest activity drifted away from the Computation Center. Bruce Franklin was eventually charged with inciting the students and was fired by the university.[50]

Charles Rosen left SRI International in 1987 to start a company focused on artificial intelligence named Machine Intelligence Corporation. Students and faculty moved easily between SRI and Stanford, most notably Nils Nilsson, who had left SRI to chair the university's department of computer science in 1985. Joint research projects spanned the two institutions, and while the trustees could no longer claim authority over SRI and the students might have had cause to feel mollified by the separation, they remained connected, and SRI was now increasingly free to engage in military projects without protest. While the reckoning had achieved some clear lines between the two organizations, not much had changed.

The field and methods of computer vision were created with the financial support of the U.S. Department of Defense and were born out of a series of military goals that directed the research and development of automated perception along narrow and well-defined lines. The discontent and protest directed toward the increasingly uncomfortable and unethical university–military partnerships of the 1960s and 1970s that resulted in the splitting of SRI International from Stanford University and CAL from Cornell University included several of the individuals involved in computer vision research programs. Frank Rosenblatt had left CAL for Cornell before the university's divestment from that institution, and by the time he took up strong positions against the Vietnam War, especially the use of advanced technology for warfare, he was no longer actively involved in machine learning research. In an oral history of his DARPA-related research, Nils Nilsson notes that this agency was the only one willing to fund the artificial intelligence and machine learning research that was intimately connected with computer vision, but without direct military application, it was no longer possible to find funding for this research.[51] Nilsson also notes the major impact of the "Lighthill Report," a 1973 report on the field of artificial intelligence to the British Science Research Council that was authored by James Lighthill, at the time the Lucasian Professor of Applied Mathematics at Cambridge University. In his report, Lighthill comments:

> Most workers in AI research and in related fields confess to a pronounced feeling of disappointment in what has been achieved in the past twenty-five years. Workers entered the field around 1950, and even around 1960, with high hopes that are very far from having been realised in 1972. In no part of the field have the discoveries made so far produced the major impact that was then promised.[52]

While Lighthill gives only glancing attention to computer vision research—he singles out as an example of one of the potentially "industrially important" practical applications the pattern-recognition task that we now call optical character recognition—his main target, especially from Nilsson's perspective, was the prospect for progress in artificial intelligence as applied to robotics.[53] Lighthill's report was highly influential on the funding of artificial intelligence research in the 1970s in the United Kingdom, but its devasting statements on the field produced repercussions that were also felt in the United States. While disappointment and discontent with the military-industrial complex certainly shifted some priorities and changed the nature and goals of smaller research projects, it was the passing of the Mansfield amendment and a change in priorities on behalf of the Department of Defense and private corporations as well as a turn away from the use of some machine learning and artificial intelligence methods to solve real-world problems that would have the largest impact on the future development of computer vision.[54]

Coda

Almost all the algorithms and methods analyzed in the previous pages are readily and freely available in some form within one of the several collections of programs, algorithms, and pretrained models known as computer vision tool kits. The presence of these historical methods in contemporary tool kits motivated, in part, the preceding analysis. At the present time, the most popular tool kit is OpenCV. This tool kit was initially created and supported by the Intel Corporation as a research project in 1999.[1] As its name implies, OpenCV is open source. It was first released under a BSD (Berkeley Software Distribution) license, then under the Apache 2 license beginning with OpenCV 4.5.0. Both licenses enable the free distribution, use, and modification of the code. Unlike the conditions of the so-called viral GNU General Public (GPL), the Apache 2 and BSD licenses permit individuals and corporations to sell modified versions of the tool kit without distributing their changes to the original code. OpenCV's selected licensing schemes make it the computer vision tool kit of choice for military organizations and other state-sponsored groups that work in partnership with private corporations. Because applications built from the code can be embedded in proprietary commercial hardware without sharing modifications or additions, OpenCV has also been widely adopted for a growing number of devices. After Intel's release of the code, the OpenCV tool kit has been maintained, expanded, and supported by a community of software developers. There are over 2,500 different algorithms included in the tool kit, although some are subject to restricted uses as a result of existing software patents. There are bindings—that is, abstract interfaces that make these algorithms easier to use and portable to different hardware and software platforms—for several major programming languages, including C++, Java, and Python. OpenCV is at present one of the core enabling technologies for computer vision. It is the foundation on which innumerable applications are built.

The widespread use of OpenCV, the existence of many common algorithms within the package, and its open-source availability make it a compelling package for critical analysis. The algorithms selected by the OpenCV developers implement many of the classic approaches to computer vision found in this book, as well as a host of more complex contemporary algorithms. In reading the code and documentation of OpenCV, one can see the persistence of many of the foundational ideas and the recurrence of earlier methods that were not as robust or reliable in their earlier implementations or with earlier hardware. OpenCV contains several specialized versions of the Hough transform discussed in chapter 4; it also has an algorithm to detect blobs, discussed in chapter 2. It includes a template-matching algorithm, and included are several methods of detecting and recognizing faces, such as Eigenfaces and the Viola-Jones framework. It has an implementation of the Perceptron in the form of a multilayer neural network.[2] OpenCV's code, documentation, and tutorial collections bundle together methods and task descriptions for picture processing, computer vision, and machine learning. In so doing, OpenCV makes apparent the ongoing connections among these activities and the difficulty of understanding computer vision isolated from the statistical methods that were necessary for their initial creation.

OpenCV proudly showcases some of what the authors present as the typical uses of their tool kit in computational vision activities around the globe:

> OpenCV's deployed uses span the range from stitching street-view images together, detecting intrusions in surveillance video in Israel, monitoring mine equipment in China, helping robots navigate and pick up objects at Willow Garage, detection of swimming pool drowning accidents in Europe, running interactive art in Spain and New York, checking runways for debris in Turkey, inspecting labels on products in factories around the world on to rapid face detection in Japan.[3]

By demonstrating the application-agnostic understanding of the utility of their product, the OpenCV package developers draw our attention to some of the most insidious forms of surveillance and control alongside some fairly benign image-processing applications. Many open-source projects, including OpenCV, are heavily used in the rapidly expanding embedded surveillance systems market and in highly customized and secretive military applications. Although some organizations have stopped using some of these surveillance tools in their applications, and some local governments have suspended the use of face recognition, the OpenCV team has yet to issue any statements about the ethics of using any of the powerful image-processing and recognition technologies provided. Documentation, tutorials, and courses on the OpenCV site provide step-by-step instructions for implementing and using a variety of these methods, including object tracking and face detection.

Among the more dramatic demonstrations of computer vision in the past years has been a series of social media–ready short videos of humanoid and doglike robotics produced by Boston Dynamics, a company that was started out of research performed at MIT, sold to Google, and then sold to Japan's SoftBank Group before being acquired by Hyundai Motor Group. Boston Dynamics takes up some of the same research goals and public-facing strategies deployed by SRI International's Shakey project. Their creation of their two-legged (Atlas) and four-legged robots (BigDog and Spot) was sponsored by DARPA and designed to "integrate with squads of Marines or Soldiers."[4] Through the wide distribution of their videos—several of their videos on YouTube have more than ten million views—which use humor, tightly orchestrated choreography between humans and machines, and dramatic pacing to produce an impressive demonstration, Boston Dynamics has placed militarized computer vision–enhanced robots back in the public consciousness.[5] Neda Atanasoski and Kalindi Vora describe the function of these videos as defanging these killer robots while at the same time working through anxieties related to autonomy: "Atlas's autonomy means that the threat of rebellion is always just under the surface in any demonstration of physical prowess and sovereign movement. To perform the limits to machine autonomy, Atlas's mobility and ability are tried and tested through pushing, prodding, and hitting that put it in its place as unfree object of human control."[6]

Boston Dynamics has also turned one of their smaller robots, Spot, into a commodity product for sale to their corporate customers.

Figure 19. Spot launch video created by Boston Dynamics, September 24, 2019.

"Spot," as they write in their advertising copy, "is an agile mobile robot that navigates terrain with unprecedented mobility, allowing you to automate routine inspection tasks and data capture safely, accurately, and frequently." Spot works at the nexus of the same fantasies and anxieties as Boston Dynamics' popular Atlas videos analyzed by Atanasoski and Vora. While Atlas is a large and menacing bipedal humanoid robot, Spot is a partly domesticated and partly autonomous four-footed guard dog. Designed for "program[able] repeatable autonomous missions," the robot turns the work site into a constantly surveilled hostile climate.[7] The launch video shows a man in an office wearing a hard hat and high-visibility vest closing his laptop and heading out while Spot comes to life, spending the evening walking and monitoring a shop floor, a construction site, and the grounds. The video ends with Spot returning to the office before the man with the hard hat closes his laptop to leave once again. As the robotic extension and threatening underside of the professional-managerial class, Spot takes over the night shift and returns faithfully to the office each morning.[8] The robot comes with a programming environment and application programming interface (API) for the Python programming language that makes use of OpenCV a core component. This tool kit operationalizes the stream of video images acquired by the device's cameras by preparing these images for classification and object recognition. The tutorial for Spot's API contains an exercise to train the device to locate objects and pick them up. A sample training activity includes the identification of a dog toy; Boston Dynamics provides a video that enables you to see what Spot sees as it moves its head and camera around a room and locates an object identified as a dog toy enclosed within a green bounding box and a confidence score. Behind many of the tasks invoked in the Spot documentation we find OpenCV, a core enabling technology for this platform, as well as updated versions of many of the algorithms and methods invoked in the previous chapters. This brief account of Boston Dynamics's commercialization of their military-industrial partnership, returns us to a number of common themes and concerns found in earlier projects, from the mobile frame of Shakey to the primacy of targeting and surveillance in object recognition tasks. The recurrence of the perspective of the machine, the OpenCV-enabled real-time display of what Spot "sees," will now motivate our return to the theory of sightless seeing.

In a talk given in 2003, German filmmaker and theorist Harun Farocki describes in precise and calculated language his view of the real-time operation of a HIL, or "hardware in the loop," simulator of a military warhead locating a target:

> The tactical warhead stores and processes the aerial photos, and the processing of the photos can be seen in green and red lines. The green lines appear to suggest something like an initial suspicion. The search-target program discovers a constellation in a picture, perhaps a part of a recognizable pattern, and stores it. The program then draws a line in the picture and searches again for an aggregate of pixels that would allow it to continue drawing that line. When the line is verified, when the outlines of a street-crossing, bridges, or power lines appear, which are registered as landmarks, the colour red is used to show that they have been verified, rather like a somewhat slow-moving mind that underlines in red a thought that seems to be correct.[9]

Farocki's talk was published as "Phantom Images," with his title using a term of art in early filmmaking: "Film recordings taken from a position that a human cannot normally occupy were called phantom shots; for example, shots from a camera that had been hung under a train." Inspired by such phantom images, Farocki created films that attempted to mimic, for viewers, the updated display of object recognition and learning from such impossible sites. It is the world as seen by a warhead that concerns Farocki. He uses his scene of simulation to revise what he previously had called "operative images," those phantom images that, he explains, "are images that do not represent an object, but rather are part of an operation."[10] Operative images would include those data stored from what might be "perhaps a part of a recognizable pattern." These are not pictures or representations of objects. Taken together with other categories of phantom images, these images form the visual apparatus of computer vision as training, test, and intermediate data as they are passed through multiple layers of typical workflows. They also fill the archives of aerial reconnaissance photographs used to get the project of computer vision off the ground. Farocki's description of and interest in the perspective of such scenes of automated computer vision returns us to the deep intimacies between computer and military vision uncovered in the previous chapters.

In recalling how Farocki made use of operative images in his Eye/Machine series, Trevor Paglen questions the value of displays like the one described above: "I couldn't quite understand why he thought these bits of visual military-industrial-complex detritus were worth paying much attention to."[11] Paglen argues that these operative images are not at all essential to the operation of computer vision and that the aesthetic use of these images is anachronistic: "Machines don't need funny animated yellow arrows and green boxes in grainy video footage to calculate trajectories or recognize moving bodies and objects. Those marks are for the benefit of humans—they're meant to show humans how a machine is seeing."[12] While Paglen's account of the operative image is a bit narrower than that found in Farocki—the operative image in his account includes, after all, not only these overlayed lines and bounding boxes but also the patterns extracted from feature detection algorithms—his naming of the operative image as "bits of visual military-industrial-complex detritus" points to this book's core issue: computer vision's function and history are made from the discarded products of training, fine-tuning, and executing computer vision. Foregrounding this otherwise off-screen activity in creative productions has an important function—not as a demonstration of the step-by-step operation of the algorithms but as an exposure of the origins of computer vision.

I'd like to return to the discussion in chapter 1 of the theoretical dimensions of sightless seeing, as theorized by Paul Virilio, and complicate the above account of the operative image and the ontology of the image in computer vision in general by considering the role these images play in recent computer vision models. Improvements in automation and robotics in the twenty-first century have moved computer vision much closer to Virilio's conception of sightless seeing, but the encoding of human perception and knowledge is still necessary and present in these technologies. Virilio's sense, already in the late 1980s, that human perception would shortly be removed or excluded from the scene of computer vision is crucially important to understanding the ontological construction of computer vision in the present. He writes:

> Once we are definitively removed from the realm of direct or indirect observation of synthetic images created *by the machine for the machine,* instrumental virtual images will be for us the equivalent of what a foreigner's mental pictures already represent: an enigma. . . . Having no graphic or videographic outputs, the automatic-perception prosthesis will function like a kind of mechanized imaginary from which, this time, we would be totally excluded.[13]

In the preceding chapters, I have shown that many of the basic computer vision operations were created in the absence of graphic output. Indeed, these techniques were created before the development of computer graphics, and many of the devices performing automatic image recognition operated without any possible graphical display. It is thus not primarily the absence of output imagery as a potential target of the human gaze that would define sightless seeing but rather the shift to an automatic framing. In the period covered by this book, that model was primarily located in SRI's Shakey the Robot. With the twenty-first-century return of neural networks to computer vision, as well as the automated extraction of complex features produced as input for other algorithmic processes, we might consider these new generated operative images as also taking part in the logics of sightless seeing.

There are, after all, many instances in which computer vision methods generate excess visual data and those situations in which images are created that are not exactly intended for viewing by human eyes. Such images are conceptually interesting within computer vision because they are not aimed at human eyes but are rather made for computer vision, even if such "vision" is an approximation of prior perceptive acts made by humans. In rendering computer vision data into image form, the data become addressable as images, a more easily understood form for humans as well as an addressable format for many image analysis algorithms. These digital objects may be created as operational placeholders, as intermediate repositories of numerical values in between automated stages of a workflow; or perhaps they are created out of a need to render data created as output from one process in image form to be used as visual input for another process; or maybe data are combined or reduced from a number of images and created in image form in order to calculate the differences among these sets of images.

With the increasing presence of computer vision applications in automated activities and in surveillance applications, the ability to interpret the decisions made from visual data becomes important. Rendering decision boundaries and categorizations made by black box algorithms or incredibly complex neural networks viewable by humans could make these operations less opaque. The creation of images from the important neurons—which is to say those elements thought to be "activated" by a coherent category of visual input—within these networks can be conceived of as a type of operative image. They become images of perception. Fabian Offert and Peter Bell borrow from W. J. T. Mitchell the highly apt concept of the metapicture to theorize the image-making activity involved in these aspects of the critical analysis of computer vision.[14] A category of image within this class that has been given some scholarly attention includes those produced by convolutional neural networks that visualize aspects of hidden layers, weights, and coefficients used for image classification decisions. Jenna Burrell has noted that because these data are not primarily designed for human viewing, it can be difficult or impossible to interpret them as images:

> What is notable is that the neural network doesn't, for example, break down handwritten digit recognition into subtasks that are readily intelligible to humans, such as identifying a horizontal bar, a closed oval shape, a diagonal line, etc. This outcome, the apparent non-pattern in these weights, arises from the very notion of computational "learning." Machine learning is applied to

> the sorts of problems for which encoding an explicit logic of decision-making functions very poorly.[15]

Burrell warns of the problem of opacity in such models as a "mismatch between mathematical procedures of machine learning algorithms and human styles of semantic interpretation."[16] This opaque dimension of operative images in contemporary sightless seeing models has also been used for creative ends. In recent years, the data learned by neural networks or highly specialized segments of these networks have been applied to artificial or noisy data and used to produce fascinating visualizations that create surrealistic or psychedelic images of cat or dog faces, which are decidedly not present in the original images. These visualizations are not so far removed from the nonintelligible, nonpattern shapes seen by Burrell in a network trained to recognize printed digits. However, rather than being unintelligible yet statistically meaningful features, they are non–statistically meaningful, highly intelligible features. Hito Steyerl calls these inceptionist images, after the Google-developed neural network that was used to produce them. She says of these: "But inceptionism is not just a digital hallucination. It is a document of an era that trains its smart phones to identify kittens, thus hardwiring truly terrifying jargons of cutesy into its means of prosumption. It demonstrates a version of corporate animism in which commodities are not only fetishes but dreamed-up, franchised chimeras. Yet these are deeply realist representations."[17] Generated by overrecognizing bits of detritus from the masses of training data collected by Google, the promise of these inception images is that in seeing them, people will be able to see as the machine sees. These are highly mechanized images, and part of their attraction is found in the seemingly endless reproduction of animal faces or other living creatures.

The most important category of the operative image, however, would be those used as input for these contemporary networks. These are operative and detritus in the sense that they, like so much of the excess and waste involved in contemporary digital culture, are discarded after the extraction of features from the images. These features alone, as we have learned, are usually not enough. The encoded perceptual knowledge of these images and the objects contained within them are attached to the extracted features.

These are the images that make sightless seeing possible. Luciana Parisi's revision of Virilio's account of sightless seeing turns on the production and extraction of knowledge from these images to question the terms of the machine–human vision binary: "Debates about the aesthetic possibilities of machine vision based on the opposition between optical human perception and non-optical automated perception seem to reinforce, instead of radically challenging, the model of metaphysical decision correlating knowledge and ocularcentrism at the core of Western philosophy."[18] A key component of Parisi's critical account of vision machines is the labor involved in labeling images for many contemporary neural network–powered computer vision algorithms. These labels, which are intended to provide a description of the primary object found in an image, are created by human viewers of the images. No small amount of labor is involved in this labeling. Parisi points to what she terms the "machine–human infrastructure" involved in these applications in order to foreground the often racialized and frequently obscured extraction of value from human labor involved in making these algorithms work: "The extracted labor of humans as service in making objects codable for machines only serves capital to extract more value from the human–machine equation of value."[19] In her essay, Parisi points to the ongoing ocularcentrism involved in computer vision and argues that the problem is not the absence of human perception from computer vision or the creation of operative images as such, but rather the ways in which human surrogate labor is being used to teach these new technologies of seeing the same old epistemologies of knowledge and vision.

The algorithms and methods of computer vision were created to encode extracted knowledge from people and images. The discourse of automaticity in the present primarily serves to obscure the values and hierarchies of knowledge encoded within these image data, operational or not. The proliferation of opaque models in which the criteria by which decisions cannot be understood works in favor of obscuring these values. The ideological value in representing overrecognized images is a return to the operation, the extraction of excess, by foreclosing the ability to see the detritus of the complex that generated this very excess. Farocki writes, "If we take an interest in pictures that are part of an operation, this is because we are weary of non-operative pictures, and weary of metalanguage."[20] The excessive repetition of mechanical operations found in the operative image is a sign of the overproduction of an ever-present computer vision that ceaselessly stitches together bits of detritus from all imaginable sensed scenes.

The experimental computer vision technologies created in the laboratories of university-affiliated research institutes and centers during the roughly twenty-year period from the late 1950s until the late 1970s have become ubiquitous in the twenty-first century. These technologies, although in altered form and now combined with faster computers and high-resolution cameras, are used in numerous everyday applications. With the initial shift toward digital photography from the film camera to the digital camera, face-detection algorithms were added that reframed the image by drawing and focusing attention on the faces found within the visual field. Smartphones have now introduced additional layers of computer vision that augment reality by altering images as soon as they are sensed by the now standard set of multiple high-resolution cameras, from the "art selfie" feature of Google's Arts & Culture app that matches a face to an artwork found in one of Google's partner museums to the profusion of Instagram and TikTok filters that can alter almost any imaginable aspect of an image, including rendering human faces in cutesy animal form. As these sample applications demonstrate, computer vision has modified and added new elements to visual culture that have changed how people see the world. In seeing the world through the products of computer vision, perception of the world outside the scope of the camera's lens has also changed. What has been added is more than a mere new photographic style or visual technique; there has been a transformation of human and machine perception. These perceptual changes are naturalized and embedded in numerous reflective and analytical practices and have added new dimensions to visual surveillance and practices of governing.

As the previous chapters have argued, the contemporary pervasive computer vision–enhanced surveillance structures are the legacy of technological solutions to mid-twentieth-century demands for visual superiority. These demands were made by the U.S. Department of Defense and answered by the emergent military-industrial complex that formed from the university and corporate partnerships developed during World War II. The goal during the initial development of these techniques was to create a visual intelligence gathering system that operated according to the logics of pattern recognition. The early algorithms and methods critiqued and interpreted genealogically in the previous chapters—the Perceptron, blob detection, template pattern matching, pictorial structures, and the Hough transform—were not consigned to the bin along with the remains of the physical machines on which they were initially created and used. They were selected for analysis because these key algorithms have survived and continue to be used in the present. The network of histories residing in these technologies are activated through everyday encounters with computer vision. These encounters are not just technical but are also found in social and psychic mechanisms and influence individual self-knowledge, habits, and even our desires.

Notes

INTRODUCTION

1. Virginia Eubanks, *Automating Inequality: How High-Tech Tools Profile, Police, and Punish the Poor* (New York: St. Martin's Press, 2017), 9–13.
2. The correlations produced across many variables learned from large amounts of training data and the opacity with which these learning algorithms operate leads to well-known problems in the interpretability of machine learning that make it difficult, if not impossible, to account for the exact variables that informed a particular decision. If such variables were made available and concrete, then individual cases might rest more on shared variables with others (e.g., home zip code) rather than a variable directly linked to the issue at hand (e.g., income). Rouvroy calls this governing practice data behaviorism. Antoinette Rouvroy, "The End(s) of Critique: Data Behaviourism versus Due Process," in *Privacy, Due Process, and the Computational Turn: The Philosophy of Law Meets the Philosophy of Technology*, ed. Mireille Hildebrandt and Katja de Vries (New York: Routledge, 2012).
3. Safiya Umoja Noble, *Algorithms of Oppression: How Search Engines Reinforce Racism* (New York: New York University Press, 2018).
4. While much of the personalization involved in surveillance capitalism operates on data, the data collected by the major corporations invoked by Zuboff—Apple, Google, Amazon, Microsoft, and Facebook—are derived from visual data and users' interactions with visual culture such as videos, games, and augmented reality. Shoshana Zuboff, *The Age of Surveillance Capitalism: The Fight for a Human Future at the New Frontier of Power* (New York: Public Affairs, 2020).
5. Following Schnepf's reading of the production and circulation of "domestic drone photographs as promoting and enabling the perpetuation of drone warfare abroad," we can understand the uptake of computer vision techniques, especially face recognition, as licensing the use of the technologies in policing and military operations. In consenting to computer vision–aided surveillance techniques, users of the systems are agreeing that these are not only effective but sensible governing technologies. J. D. Schnepf, "Domestic Aerial Photography

in the Era of Drone Warfare," *Modern Fiction Studies* 63, no. 2 (2017): 270–87, 272, https://doi.org/10.1353/mfs.2017.0022.

6. Jonathan Crary, "Techniques of the Observer," *October* 45 (1988): 3–35, 35, https://doi.org/10.2307/779041.
7. In relation to the larger cultural and historical frameworks of degradation or "denigration" of human vision that make this feature of computer vision seem less aberrant and more of the same, see Martin Jay, *Downcast Eyes: The Denigration of Vision in Twentieth Century French Thought* (Berkeley: University of California Press, 1993).
8. The lack of display, and hence the concept of a screen, as a required component of computer vision complicates some of the analytical framework that otherwise might be imported from film theory. Nonetheless, the question of the relation between gaze and look between spectator and spectacle are helpful in understanding the import of the historical trajectories and ontological shifts in computer vision that have reconceptualized the field of vision. For a useful psychoanalytical account of these concepts, see Kaja Silverman, *The Threshold of the Visible World* (New York: Routledge, 1996), 125–61. Also germane to these distinctions is Denson's concept of discorrelated images—that is, shots used in contemporary film and video game images that question and confuse subjective and objective views. Denson argues that these function to "dismantle the rational orderings of time and space that served, conventionally, to correlate spectatorial subjectivity with cinematic images." Shane Denson, *Discorrelated Images* (Durham, N.C.: Duke University Press, 2020), 8–9.
9. Donna Haraway, "Situated Knowledges: The Science Question in Feminism and the Privilege of Partial Perspective," *Feminist Studies* 14, no. 3 (1988): 575–99, 581.
10. Haraway, "Situated Knowledges," 581.
11. Haraway, "Situated Knowledges," 581.
12. The critical inquiry into the interaction between the digital tool and the researcher making use of the tool has been termed tool criticism. See Karin van Es, Maranke Wieringa, and Mirko Tobias Shäfer, "Tool Criticism: From Digital Methods to Digital Methodology," in *Proceedings of the 2nd International Conference on Web Studies* (New York: Association for Computing Machinery, 2018), 24–27, https://doi.org/10.1145/3240431.3240436.
13. An important book taking a mostly phenomenological perspective on algorithms and algorithmic culture is Ed Finn, *What Algorithms Want: Imagination in the Age of the Computer* (Cambridge, Mass.: MIT Press, 2017). The question of access to algorithms and code, especially given the recent prevalence of black box applications, has been important in developing interpretive methods for technology. For a theoretical

account of the possibilities of phenomenology for algorithm studies, see Johannes Paßmann and Asher Boersma, "Unknowing Algorithms: On Transparency of Unopenable Black Boxes," in *The Datafied Society*, ed. Mirko Tobias Schäfer and Karin van Es (Amsterdam: Amsterdam University Press, 2017), 139–46. Annany and Crawford critique what many offer up as the "transparency ideal," the assumption that access to code and algorithms is all that is necessary to understand technology, by refocusing attention on the systems invoked by Hughes. Mike Ananny and Kate Crawford, "Seeing without Knowing: Limitations of the Transparency Ideal and Its Application to Algorithmic Accountability," *New Media and Society* 20, no. 3 (2018): 973–89, https://doi.org/10.1177/1461444816676645.

14. Hughes has been influential in this interpretive framework. He turns from specific artifacts to the technological systems in which these artifacts are embedded, which he glosses as "organizations, such as manufacturing firms, utility companies, and investment banks, and they incorporate components usually labeled scientific, such as books, articles, and university teaching and research programs. Legislative artifacts, such as regulatory laws, can also be part of technological systems." Thomas P. Hughes, "The Evolution of Large Technological Systems," in *The Social Construction of Technological Systems: New Directions in the Sociology and History of Technology*, ed. Wiebe E. Bijker, Thomas P. Hughes, and Trevor Pinch (Cambridge, Mass.: MIT Press, 2012), 45–76, 45.
15. Sample studies and critiques include the following: Ruha Benjamin, *Race After Technology: Abolitionist Tools for the New Jim Code* (Medford, Mass.: Polity, 2019); Sun-ha Hong, *Technologies of Speculation: The Limits of Knowledge in a Data-Driven Society* (New York: New York University Press, 2020); Emily M. Bender, Timnit Gebru, Angelina McMillan-Major, and Shmargaret Shmitchell, "On the Dangers of Stochastic Parrots: Can Language Models Be Too Big?," in *Proceedings of the 2021 ACM Conference on Fairness, Accountability, and Transparency* (New York: Association for Computing Memory [ACM], 2021), 610–23, https://doi.org/10.1145/3442188.3445922.
16. The growing concern with the public use of face recognition technology can be registered in recent activism and political activity asking for regulating or banning these methods. Consider the vote by San Francisco's board of supervisors to block the use of face recognition by the police department. Kate Conger, Richard Fausset, and Serge F. Kovaleski, "San Francisco Bans Facial Recognition Technology," *New York Times*, May 14, 2019. Facebook, which has long used face recognition to label newly uploaded photos and to suggest new connections, claims to have halted their use of the technology and deleted the data

generated by these algorithms in 2021. Kashmir Hill and Ryan Mac, "Facebook, Citing Societal Concerns, Plans to Shut Down Facial Recognition System," *New York Times,* November 2, 2021.

17. Sayers is one of the few scholars who has considered the history of computer vision from a humanist perspective in his speculative work with these technologies. Jentery Sayers, "Bringing Trouvé to Light: Speculative Computer Vision and Media History," in *Seeing the Past with Computers: Experiments with Augmented Reality and Computer Vision for History,* ed. Kevin Kee and Timothy Compeau (Ann Arbor: University of Michigan Press, 2019), 32–49. See also Adrian Mackenzie, "Simulate, Optimise, Partition: Algorithmic Diagrams of Pattern Recognition from 1953 Onwards," in *Cold War Legacies: Legacy, Theory, Aesthetics,* ed. John Beck and Ryan Bishop (Edinburgh: Edinburgh University Press, 2016), 50–69; Lev Manovich, "The Automation of Sight: From Photography to Computer Vision," in *Electronic Culture: Technology and Visual Representation,* ed. Timothy Druckrey (New York: Aperture Foundation, 1996), 229–39.
18. One of the major ideological elements of computer technology and the computer industry is the presentation of the history of technology as a progressive narrative in which ever newer, ever better technologies replace and obsolete what has come before. This progress myth assumes that the horizon of futurity brought about by new computing technologies will necessarily improve the world. Winner calls this belief in the progressive and revolutionary capacity of computing systems "mythinformation." Langdon Winner, *The Whale and the Reactor: A Search for Limits in an Age of High Technology* (Chicago: University of Chicago Press, 1988).
19. Computational vision and machine learning in military applications may well render all warfare virtual, as sensed scenes are transformed into data and processed. This is the now-familiar claim made by Jean Baudrillard, *The Gulf War Did Not Take Place,* trans. Paul Patton (Bloomington: Indiana University Press, 1995). A Baudrillard-influenced account of computer vision and artificial intelligence used in intelligence analysis as part of warfare can be found in Manuel De Landa, *War in the Age of Intelligent Machines* (New York: Zone Books, 1991). The tensions between embedded computation and virtual warfare found in twentieth-century computer vision research inform today's discourse. In many reports on the use of drones, the military has made an effort to stress the remote but human operation of these weapons-capable flying drones. In reality, they are essentially flying computers with advanced onboard image-processing capabilities. On the more recent history of the use of robotics in the military and the development of military drones, see P. W. Singer, *Wired for War: The Robotics Revolution and Conflict in the Twenty-first Century* (New York: Penguin, 2009).

20. Brian Massumi, *Ontopower: War, Powers, and the State of Perception* (Durham, N.C.: Duke University Press, 2015), 65–67.
21. Paul E. Ceruzzi, *A History of Modern Computing* (Cambridge, Mass.: MIT Press, 1998), 143.
22. From the beginning of the field to the present, many specialized computing devices have been created for computer vision. Recently vendors have created AI accelerators to offload some computational tasks that are especially well suited to computer vision. Google's tensor processing unit (TPU) is designed for such uses and has been marketed to scientific computing users and hobbyists; it has also been added as a "Pixel neural core" to Pixel 4 smartphones. Norman Jouppi, Cliff Young, Nishant Patil, and David Patterson, "Motivation for and Evaluation of the First Tensor Processing Unit," *IEEE Micro* 38, no. 3 (2018): 10–19, https://doi.org/10.1109/MM.2018.032271057. OpenCV has codeveloped with the Luxonis Holding Corporation a hardware product called OAK-D that provides multiple cameras for stereovision and hardware acceleration to offload neural network–based algorithms for object detection in three-dimensional rather than pixel (image) space (https://docs.luxonis.com/en/latest/).
23. Ceruzzi, *History of Modern Computing*, 153–54.
24. An example is this accounting of the performance of an imagined algorithm: "Assuming 10–3 s per evaluation, we would require 1.5 × 105 s or approximately two days of computation time. It is thus obvious that a more effective technique is required. . . . The computational feasibility of the DP [dynamic programming] approach depends on storage and computing time requirements." Martin A. Fischler and Robert A. Elschlager, "The Representation and Matching of Pictorial Structures," *IEEE Transactions on Computers* C-22, no. 1 (1973): 67–92, 70, https://doi.org/10.1109/T-C.1973.223602.
25. I borrow the phrase "overrapid historicization" from Žižek, who uses this concept in concert with overrapid universalization to find a way through an impasse within critical approaches with culture. Slavoj Žižek, *The Sublime Object of Ideology* (New York: Verso, 1989).
26. On Amazon's Rekognition program and its marketing of this software to police departments, see Nick Wingfield, "Amazon Pushes Facial Recognition to Police. Critics See Surveillance Risk," *New York Times*, May 22, 2018. An excellent account of the history of surveillance of Black people is Simone Browne, *Dark Matters: On the Surveillance of Blackness* (Durham, N.C.: Duke University Press, 2015).
27. The best source for a technical history of these fields is Nils J. Nilsson, *The Quest for Artificial Intelligence: A History of Ideas and Achievements* (Cambridge: Cambridge University Press, 2010).
28. Richard O. Duda and Peter Hart, *Pattern Classification and Scene Analysis* (New York: Wiley, 1973), vii.

29. Duda and Hart, *Pattern Classification,* 1.
30. Jay David Bolter and Richard Gruisin, *Remediation: Understanding New Media* (Cambridge, Mass.: MIT Press, 1999), 20.
31. For an analysis of this aspect of algorithms, see Alexander Galloway, *Gaming: Essays on Algorithmic Culture* (Minneapolis: University of Minnesota Press, 2006); Ted Striphas, "Algorithmic Culture," *European Journal of Cultural Studies* 18, no. 4–5 (2015): 395–412, https://doi.org/10.1177/1367549415577392. See also Rob Kitchin, "Thinking Critically About and Researching Algorithms," *Information, Communication and Society* 20, no. 1 (2016): 1–16, https://doi.org/10.1080/1369118X.2016.1154087.
32. Nick Seaver, "Algorithms as Culture: Some Tactics for the Ethnography of Algorithmic Systems," *Big Data and Society* 4, no. 2 (2017): 1–12, https://doi.org/10.1177/2053951717738104.
33. Yanni Alexander Loukissas, *All Data Are Local: Thinking Critically in Data-Driven Society* (Cambridge, Mass.: MIT Press, 2019), 117.
34. Peter Galison, "The Ontology of the Enemy: Norbert Wiener and the Cybernetic Vision," *Critical Inquiry* 21, no. 1 (1994): 228–66, 264, https://doi.org/10.1086/448747.
35. Orit Halpern, *Beautiful Data: A History of Vision and Reason since 1945* (Durham, N.C.: Duke University Press, 2015), 201–2.
36. The Perceptron will generally be capitalized because despite the wishes of Rosenblatt, it names a specific invention rather than providing a "generic name for a variety of theoretical nerve nets." He jokingly compares capitalizing the "P" in Perceptron to the "E" in electron. Frank Rosenblatt, *Principles of Neurodynamics: Perceptrons and the Theory of Brain Mechanisms* (Washington, D.C.: Spartan Books, 1962), v.
37. Michel Foucault, "Nietzsche, Genealogy, History," in *Language, Counter-memory, Practice: Selected Essays and Interviews,* ed. D. F. Bouchard (Ithaca, N.Y.: Cornell University Press, 1977), 139.

1. COMPUTER VISION

1. Aleksandr Georgievich Arkadev and Emmanuil Markovich Braverman, *Computers and Pattern Recognition,* trans. W. Turski and J. D. Cowan (Washington, D.C.: Thompson Book, 1967).
2. Braverman applied his machine learning work to several different fields, including cancer research and mineral classification. He also produced theoretical work on problems in cybernetics and the engineering of "control" in terms of behavior to be applied to Soviet economics. See Olessia Kirtchik, "From Pattern Recognition to Economic Disequilibrium," *History of Political Economy* 51, no. 1 (2019): 180–203, https://doi.org/10.1215/00182702-7903288.

3. Lev Rozonoer, Boris Mirkin, and Ilya Muchnik, eds., *Braverman Readings in Machine Learning: Key Ideas from Inception to Current State* (Cham, Switzerland: Springer International, 2018).
4. A. Klinger, J. K. Aggarwal, N. J. George, R. M. Haralick, T. S. Huang, and O. J. Tretiak, "Soviet Image Pattern Recognition Research," technical report prepared by Science Applications International (Washington, D.C.: U.S. Department of Energy, Office of Scientific and Technical Information, December 1989), https://doi.org/10.2172/6764408.
5. ARPA changed its name to DARPA before changing back to ARPA and then DARPA, the present name of the organization, once again. For an example of the international scope of early computer vision featuring contributions from Italian, Indian, Swiss, German, Austrian, English, Canadian, and American scholars, see Antonio Grasselli, ed., *Automatic Interpretation and Classification of Images: A NATO Advanced Study Institute* (New York: Academic Press, 1969). This volume collects talks given at a NATO summer school under the published volume's title in Pisa-Tirrenia, Italy, from August 26 to September 6, 1978.
6. This is especially evident in the control room designed for the Tobermory Perceptron. George Nagy, "System and Circuit Design for the Tobermory Perceptron," Cognitive Systems Research Program Report 5 (Ithaca, N.Y.: Cornell University, 1963), 4–5.
7. The rapid growth of image-based social media applications has made these methods a sign of distinction as well as a necessity because these application vendors and platform producers need systems and software capable of addressing and manipulating many billions of images.
8. Yuk Hui, *On the Existence of Digital Objects* (Minneapolis: University of Minnesota Press, 2016), 1.
9. Hui, *On the Existence of Digital Objects,* 72.
10. Azriel Rosenfeld, "Progress in Picture Processing: 1969–71," *Computing Surveys* 5, no. 2 (1973): 81–102, 81.
11. Manovich, "Automation of Sight," 233. Manovich's essay was written before the full return to computer vision of neural networks, which operate without a specified formal model of either space or visual features. Restoring the Perceptron to the history of computer vision and connecting it to its descendants in the range of contemporary neural network architectures shows that although computer graphics and computer vision are coarticulated, they are not necessarily joined in vision or perspective.
12. Jacob Gaboury, *Image Objects: An Archeology of Computer Graphics* (Cambridge, Mass.: MIT Press, 2021), 23.
13. Azriel Rosenfeld, *Picture Processing by Computer* (New York: Academic Press, 1969), 4.
14. Ian Goodfellow, Jean Pouget-Abadie, Mehdi Mirza, Bing Xu, David

Warde-Farley, Sherjil Ozair, Aaron Courville, and Yoshua Bengio, "Generative Adversarial Networks," *Communications of the ACM* 63, no. 11 (2020): 139–44, https://doi.org/10.1145/3422622.

15. André Bazin, "The Ontology of the Photographic Image," trans. Hugh Gray, *Film Quarterly* 13, no. 4. (1960): 4–9, 8.
16. Susan Sontag, *On Photography* (London: Penguin, 2008), 69.
17. Bazin, "Ontology of the Photographic Image," 7.
18. A useful philosophical account of the representative aspect of images as being located in the relation among structural features rather than in the perception of images is John V. Kulvicki, *On Images: Their Structure and Content* (Oxford: Oxford University Press, 2006).
19. Jason Salavon, *Jason Salavon: Brainstem Still Life* (Bloomington: Indiana University School of Fine Arts Gallery, 2004).
20. On the possibility of generality in photography and the strategies deployed by photographers in relation to such meaning-making operations, see Roland Barthes, *Camera Lucida: Reflections on Photography*, trans. Richard Howard (New York: Vintage Classics, 2000), 34–38.
21. Paul Virilio, *The Vision Machine*, trans. Julie Rose (Bloomington: Indiana University Press, 1994), 59.
22. Virilio, *Vision Machine*, 59–60.
23. MIT's Project MAC was established in 1963 and shared staff and research projects with the artificial intelligence group. MAC, according to Edwards, "stood variously for Man and Computer, Machine-Aided Cognition, or Multi-Access Computing." Paul N. Edwards, *The Closed World: Computers and the Politics of Discourse in Cold War America* (Cambridge, Mass.: MIT Press, 1996), 269.
24. Seymour Papert, "The Summer Vision Project," MIT Artificial Intelligence Memo 104-1, July 7, 1966.
25. Lawrence G. Roberts, "Machine Perception of Three-Dimensional Solids" (PhD diss., Massachusetts Institute of Technology, 1963). Gaboury locates Roberts's dissertation within the nascent field of computer vision, but he also sees it as an important step toward the development of three-dimensional images within the field of computer graphics. Gaboury, *Image Objects*, 39–44.
26. Azriel Rosenfeld, "Computer Vision: Basic Principles," *Proceedings of the IEEE* 76, no. 8 (1988): 863–68.
27. Patrick Henry Winston, *The Psychology of Computer Vision* (New York: McGraw-Hill, 1975); Allen R. Hanson and Edward M. Riseman, *Computer Vision Systems* (New York: Academic Press, 1978).
28. Ballard and Brown define computer vision as "the enterprise of automating and integrating a wide range of processes and representations used for vision perception. It includes as parts many techniques that are useful by themselves, such as *image processing* (transforming, encoding, and transmitting images) and *statistical pattern classifica-*

tion (statistical decision theory applied to general patterns, visual or otherwise)." Dana H. Ballard and Christopher M. Brown, *Computer Vision* (Englewood Cliffs, N.J.: Prentice-Hall, 1982), 2.
29. Fischler and Firschein, *Intelligence,* v.
30. Norbert Wiener, *Cybernetics, or Control and Communication in the Animal and the Machine* (Cambridge, Mass.: MIT Press, 1948).
31. Ballard and Brown, *Computer Vision,* 5.
32. Arkadev and Braverman, *Computers and Pattern Recognition,* 10.
33. Winston, *Psychology of Computer Vision,* 2.
34. Fischler and Firschein, *Intelligence,* 239.
35. Virilio notes the renewed interest in human perception brought about through developments in computer vision and adjacent research: "It was not until the 60s and work on optoelectronics and computer graphics that people began to take a fresh look at the psychology of visual perception, notably in the United States." Virilio, *Vision Machine,* 60.
36. Arkadev and Braverman, *Computers and Pattern Recognition,* 13.
37. Arkadev and Braverman, *Computers and Pattern Recognition,* 13.
38. OpenCV's matchTemplate() function provides a contemporary version of this technique. OpenCV, "Template Matching," https://docs.opencv.org/4.5.0/d4/dc6/tutorial_py_template_matching.html. See chapter 3 for a more thorough discussion of template matching and the creation of the pictorial structures method for pattern detection.
39. Mark C. Marino, "Reading Culture through Code," in *Routledge Companion to Media Studies and Digital Humanities,* ed. Jentery Sayers (New York: Routledge, 2018), 472–82, 473.
40. Gillespie defines an algorithm as a model that provides "the formalization of a problem and its goal, articulated in computational terms." Tarleton Gillespie, "Algorithm," in *Digital Keywords,* ed. Benjamin Peters (Princeton, N.J.: Princeton University Press, 2016), 18–30, 19.
41. Adrian MacKenzie and Anna Munster, "Platform Seeing: Image Ensembles and Their Invisualities," *Theory, Culture, and Society* 36, no. 5 (2019): 3–22, https://doi.org/10.1177/0263276419847508.
42. David Golumbia, *The Cultural Logic of Computation* (Cambridge, Mass.: Harvard University Press, 2009), 4.
43. Tara McPherson, "Why Are the Digital Humanities So White? or, Thinking the Histories of Race and Computation," in *Debates in the Digital Humanities,* ed. Matthew K. Gold (Minneapolis: University of Minnesota Press, 2012), 139–60.
44. McPherson, "Digital Humanities So White," 144.
45. Louise Amoore, *Cloud Ethics: Algorithms and the Attributes of Ourselves and Others* (Durham, N.C.: Duke University Press, 2020), 7–8.
46. Ian Bogost, *Unit Operations: An Approach to Video Game Criticism* (Cambridge, Mass.: MIT Press, 2008), 7.

47. W. S. Holmes, "Automatic Photointerpretation and Target Location," *Proceedings of the IEEE* 54, no. 12 (1966): 1679–86, https://doi.org/10.1109/PROC.1966.5249.
48. Holmes, "Automatic Photointerpretation," 1680.
49. Duda and Hart, *Pattern Classification*, 2.
50. Duda and Hart, *Pattern Classification*, 4.
51. Duda and Hart, *Pattern Classification*, 2–3.
52. Lev Manovich, *Cultural Analytics* (Cambridge, Mass.: MIT Press, 2020), 140.

2. INVENTING MACHINE LEARNING WITH THE PERCEPTRON

1. For more on the Dartmouth conference and its legacy, see James Moor, "The Dartmouth College Artificial Intelligence Conference: The Next Fifty Years," *AI Magazine* 27, no. 4 (2006): 87–89, https://doi.org/10.1609/aimag.v27i4.1911. The workshop proposal, "A Proposal for the Dartmouth Summer Research Project on Artificial Intelligence," was authored by John McCarthy, Marvin Minsky, Nathaniel Rochester, and Claude Shannon and dated August 31, 1955.
2. Fisher's linear discriminant was described in the mid-1930s. Ronald A. Fisher, "The Use of Multiple Measurements in Taxonomic Problems," *Annals of Eugenics* 7, no. 2 (1936): 179–88, https://doi.org/10.1111/j.1469-1809.1936.tb02137.x. Fix and Hodges published their research on the nearest-neighbor rule for nonparametric linear discrimination as a technical report in 1951. Evelyn Fix and Joseph L. Hodges, "Nonparametric Discrimination: Consistency Properties," Report 4, Project 21-49-004 (Randolph Field, Tex.: USAF School of Aviation Medicine, 1951).
3. Rosenblatt mentions some results of a simulated cat model derived from "the organization of the visual system of the cat," which was most likely based on Donald O. Hebb's model, on a computer at New York University in a report from the Cognitive Systems Research Program on project RR 003-08-01 to the Office of Navy Research. Information Systems Summaries (ONR Report ACR-81, July 1963).
4. The Curtiss-Wright Aircraft Corporation was long associated with military aircraft and technology. Their facility in Lockland, Ohio, was responsible for producing defective aircraft engines during World War II and served as the inspiration for Arthur Miller's play *All My Sons* (1946).
5. This dedication appears on a large metal plaque commemorating the gift of the Curtiss-Wright Aircraft Corporation to Cornell University (https://www.calspan.com/company/history/1940s/).
6. Some of this corporate history is given by the Calspan Corporation (https://www.calspan.com/company/history/).

7. "Dr. Frank Rosenblatt Dies at 43; Taught Neurobiology at Cornell," *New York Times*, July 13, 1971.
8. Frank Rosenblatt, "The k-Coefficient: Design and Trial Application of a New Technique for Multivariate Analysis" (PhD diss., Cornell University, 1956).
9. Rosenblatt, "k-Coefficient," 79.
10. Rosenblatt, "k-Coefficient," 91, 111.
11. The EPAC's design and function is provided in appendix G of Rosenblatt, "k-Coefficient," 205–10.
12. In December 1953, *Cornell Alumni Magazine* had a small feature on Frank Rosenblatt and his "idiot brain." The article was accompanied by a photograph of Rosenblatt demonstrating the reading of survey data by the EPAC system installed in Morrill Hall.
13. Frank Rosenblatt, "The Perceptron: A Perceiving and Recognizing Automaton (Project PARA)," report 85-460-1 (Buffalo, N.Y.: Cornell Aeronautical Laboratory, 1957), 3.
14. See "-tron" suffix, OED Online, https://www.oed.com/.
15. Rosenblatt, "Perceptron: Project PARA," 15.
16. Frank Rosenblatt, "The Perceptron: A Probabilistic Model for Information Storage and Organization in the Brain," *Psychological Review* 65, no 6. (1958): 386–408, 386, https://doi.org/10.1037/h0042519.
17. For more on statistical separability and the Perceptron, see Frank Rosenblatt, "Two Theorems of Statistical Separability in the Perceptron," report VG-1196-G-2 (Buffalo, N.Y.: Cornell Aeronautical Laboratory, 1958).
18. Warren S. McCulloch and Walter S. Pitts, "A Logical Calculus of the Ideas Immanent in Nervous Activity," *Bulletin of Mathematical Biophysics* 5 (1943): 115–33, https://doi.org/10.1007/BF02478259.
19. McCulloch and Pitts, "Logical Calculus," 129.
20. McCulloch and Pitts, "Logical Calculus," 131.
21. Rosenblatt, *Principles of Neurodynamics*, 13.
22. Donald O. Hebb, *The Organization of Behavior: A Neuropsychological Theory* (1949; reprint, New York: Taylor & Francis, 2002), 18.
23. In 1956, researchers at IBM implemented the Hebb model in what they called "cell assemblies" from a simulated unorganized neural network on a IBM 704 computer. N. Rochester, J. Holland, L. Haibt, and W. Duda, "Tests on a Cell Assembly Theory of the Action of the Brain, Using a Large Digital Computer," *IRE Transactions on Information Theory* 2, no. 3 (1956): 80–93, https://doi.org/10.1109/TIT.1956.1056810.
24. Jerome Y. Lettvin, Humberto R. Maturana, Warren S. McCulloch, and Walter H. Pitts, "What the Frog's Eye Tells the Frog's Brain," *Proceedings of the IRE* 47, no. 11 (1959): 1940–51.
25. On the implications of this paper for theories of embodiment and perception, see Neda Atanasoski and Kalidini Vora, *Surrogate Humanity:*

Race, Robots, and the Politics of Technological Futures (Durham, N.C.: Duke University Press, 2019), 129–33. Wilson's account of the absence of affect in Pitts's work provides additional historical context for this paper and what she terms the "homosocial" milieu of this group of collaborators. Elizabeth A. Wilson, *Affect and Artificial Intelligence* (Seattle: University of Washington Press, 2010), 109–32, 126.

26. Lettvin et al., "Frog's Eye," 1950.
27. Rosenblatt, "Perceptron: Project PARA," 3.
28. Rosenblatt, "Perceptron: Probabilistic Model," 395.
29. Edwards, *Closed World*, 10.
30. Among other places, proof of the convergence theorem is found in Rosenblatt, *Principles of Neurodynamics*.
31. Rosenblatt, "Perceptron: Project PARA," ii.
32. John C. Hay, Ben E. Lynch, and David R. Smith, "MARK I Perceptron Operators' Manual (Project PARA)," report VG-11196-G-5 (Buffalo, N.Y.: Cornell Aeronautical Laboratory, 1960), 1.
33. Albert E. Murray, "Perceptron Applicability to Photointerpretation," phase 1 report for Project PICS, report VE-1446 G-1 (Buffalo, N.Y.: Cornell Aeronautical Laboratory, 1960).
34. "Neural Network," National Museum of American History, https://americanhistory.si.edu/collections/search/object/nmah_334414.
35. Albert Maurel Uttley, "The Design of Conditional Probability Computers," *Information and Control* 2 (1959): 1–24, 1, https://doi.org/10.1016/S0019-9958(59)90058-0.
36. Uttley, "Design of Conditional Probability Computers," 9.
37. Alfred E. Brain, George Forsen, David Hall, and Charles Rosen, "A Large, Self-Contained Learning Machine," in *Proceedings of the Western Electronic Show and Convention*, San Francisco, Calif., August 20–23, 1963, C-1.
38. Antonio Borsellino and Augusto Gamba, "An Outline of a Mathematical Theory of PAPA," *Il Nuovo Cimento* 20, no. 2 (1961): 221–31.
39. Augusto Gamba, "The Papistor: An Optical PAPA Device," *Il Nuovo Cimento* 26, no. 3 (1962): 371–73.
40. George Nagy, "A Survey of Analog Memory Devices," *IEEE Transactions on Electronic Computers* EC-12, no. 4 (1963): 388–93.
41. George Nagy, "Analogue Memory Mechanisms for Neural Nets" (PhD diss., Cornell University, 1962).
42. George Nagy, "System and Circuit Designs for the Tobermory Perceptron (Preliminary Report on Phase 1)," Cornell University Cognitive Systems Research Program, September 1, 1963. Nagy cites Saki's story as H. H. Munro, *The Chronicles of Clovis* (London: Bodley Head, 1911).
43. Rosenblatt, "Perceptron: Project PARA," 3.
44. Frank Rosenblatt, "Perceptron Simulation Experiments," *Proceedings*

of the IRE 48, no. 3 (1960): 301–9, https://doi.org/10.1109/JRPROC.1960.287598.

45. W. S. Holmes, H. R. Leland, G. E. Richmond, and M. G. Spooner, "Status and Planning Report on Perceptron Applicability to Automatic Photo Interpretation," report VE-1446-G-2 (Buffalo, N.Y.: Cornell Aeronautical Laboratory, 1961), 5.
46. *The Machine That Changed the World,* episode 4, "The Thinking Machine," written and directed by Nancy Linde, WGBH, January 1992.
47. Charles A. Rosen, "Pattern Classification by Adaptive Machines," *Science* 156, no. 3771 (1967): 38–44, 39, https://doi.org/10.1126/science.156.3771.38.
48. Rosen, "Pattern Classification," 39.
49. Rosen, "Pattern Classification," 39.
50. Albert E. Murray, "Perceptron Applications in Photo Interpretation," *Photogrammetric Engineering* 27, no. 4 (1961): 627–37.
51. Murray, "Perceptron Applications," 628.
52. Murray, "Perceptron Applications," 3.
53. Holmes, "Automatic Photointerpretation," 1679.
54. W. S. Holmes, H. R. Leland, and G. E. Richmond, "Design of a Photo Interpretation Automaton," in *Proceedings of the December 4–6, 1962, Fall Joint Computer Conference on Computers in the Space Age* (New York: Association for Computing Memory [ACM] Press, 1962), 27–35, https://dl.acm.org/doi/10.1145/1461518.1461521.
55. Holmes, Leland, and Richmond, "Design of a Photo Interpretation Automaton," 27.
56. Holmes, Leland, and Richmond, "Design of a Photo Interpretation Automaton," 28.
57. Holmes's work was published several years before the Mansfield congressional amendment that required ARPA/DARPA-supported projects to have a direct military application.
58. Holmes, "Automatic Photointerpretation," 1681.
59. Holmes, "Automatic Photointerpretation," 1671.
60. Thinking Machines Corporation's Connection Machine, a series of parallel computers for high-performance computing created in the 1980s and 1990s that were based on the neural network concept and that had the financial support of DARPA, was a commercial failure for similar reasons. Other, similar systems were developed at the same time that were also not viable in the commercial computer market. Hillis, inventor of the Connection Machine and CEO of the Thinking Machines Corporation, cites the Perceptron as the precursor parallel learning machine to his own device. W. Daniel Hillis, *The Connection Machine* (Cambridge, Mass.: MIT Press, 1992), 161. Gary A. Taubes, "The Rise and Fall of Thinking Machines," *Inc.,* September 15, 1989, https://www.inc.com/magazine/19950915/2622.html.

61. Harding Mason, D. Stewart, and Brendan Gill, “Rival,” *New Yorker,* December 6, 1958, 44–45, 45, https://www.newyorker.com/magazine/1958/12/06/rival-2.
62. “New Navy Device Learns by Doing: Psychologist Shows Embryo of Computer Designed to Read and Grow Wiser,” *New York Times,* July 8, 1958.
63. Marvin Minsky, “Steps toward Artificial Intelligence,” *Proceedings of the IRE* 49, no. 1 (1961): 8–30.
64. Minsky, “Steps toward Artificial Intelligence,” 15.
65. Jerry Kaplan, *Artificial Intelligence: What Everyone Needs to Know* (Oxford: Oxford University Press, 2016), 33–34.
66. Rosenblatt, *Principles of Neurodynamics,* v–vi.
67. For a cogent summary and interpretation of the arguments between Rosenblatt and Minsky, see Mikel Olazaran, “A Sociological Study of the Official History of the Perceptrons Controversy,” *Social Studies of Science* 26, no. 3 (1996): 611–59.
68. Léon Bottou, foreword to *Perceptrons: An Introduction to Computational Geometry,* reissue of 1988 expanded ed. (Cambridge, Mass.: MIT Press, 2017), vii.
69. Cornell University’s obituary for Rosenblatt briefly recounts his career and major research projects. Stephen T. Emlen, Howard C. Howland, and Richard D. O’Brien, “Frank Rosenblatt: July 11, 1928–July 11, 1971,” Cornell University faculty memorial statement, https://ecommons.cornell.edu/bitstream/handle/1813/18965/Rosenblatt_Frank_1971.pdf.
70. “Frank Rosenblatt: Professor, Inventor of ‘Perceptron,’” *Washington Post,* July 14, 1971.
71. *Congressional Record* 117, no. 21 (July 28, 1971): H27716.
72. Rosenblatt coauthored one such study with Rodman “Rod” Miller, a Cornell University undergraduate student who lived with several other students and others in Rosenblatt’s Brooktondale, New York, home at the time of his death. Frank Rosenblatt and Rodman G. Miller, “Behavioral Assay Procedures for Transfer of Learned Behavior by Brain Extracts,” *Proceedings of the National Academy of Sciences of the United States of America* 56, no. 5 (1966): 1423–30.
73. Ralph Littauer and Norman Uphoff, eds., *The Air War in Indochina,* rev. ed. (Boston: Beacon, 1972).

3. DESCRIBING PICTURES

1. On the U.S. military’s imaginings of an electronic battlefield, see Paul Dickson, *The Electronic Battlefield* (Bloomington: Indiana University Press, 1976); Jon R. Lindsay, *Information Technology and Military Power* (Ithaca, N.Y.: Cornell University Press, 2020), 15–20.

2. Minsky, "Steps toward Artificial Intelligence," 15.
3. Richard F. Lyon, "A Brief History of 'Pixel'" (paper presented at the IS&T/SPIE Symposium on Electronic Imaging, San Jose, California, January 15–19, 2006), https://doi.org/10.1117/12.644941.
4. In their exploration of ImageNet, an important computer vision data set, Crawford and Paglen write of an "archeology of datasets" that involves "digging through the material layers, cataloguing the principles and values by which something was constructed, and analyzing what normative patterns of life were assumed, supported, and reproduced." Kate Crawford and Trevor Paglen, "Excavating AI: The Politics of Training Sets for Machine Learning," September 19, 2019, https://excavating.ai.
5. Joy Buolamwini and Timnit Gebru, "Gender Shades: Intersectional Accuracy Disparities in Commercial Gender Classification," *PMLR* 81 (2018): 77–91.
6. The social construction of technology thesis provides us with a useful frame for understanding the social significance of technologies that are frequently presented as value-free by their developers and advocates. The canonical account of this argument is Langdon Winner, "Do Artifacts Have Politics?" *Daedalus* 109, no. 1 (1980): 121–36.
7. On the development of the object templates and models approach to computer vision, see Nilsson, *Quest for Artificial Intelligence,* 263–67.
8. Microsoft, on November 8, 2021, announced this automatic captioning feature, providing some detail on the training of the tool and its performance on a benchmark called "nocaps": "Image captioning is a core challenge in the discipline of computer vision, one that requires an AI system to understand and describe the salient content, or action, in an image, explained Lijuan Wang, a principal research manager in Microsoft's research lab in Redmond. . . . Datasets of images with word tags instead of full captions are more efficient to create, which allowed Wang's team to feed lots of data into their model. The approach imbued the model with what the team calls a visual vocabulary" (https://blogs.microsoft.com/ai/azure-image-captioning/).
9. Martin A. Fischler, "Machine Perception and Description of Pictorial Data," in *Proceedings of the 1st International Joint Conference on Artificial Intelligence* (San Francisco, Calif.: Morgan Kaufmann, 1969), 629.
10. Minsky, in the conclusion of his state-of-the-field essay on artificial intelligence, suggests that the development of time sharing, with multiple people using a computer at the same time, as well as other techniques to enable real-time human–computer interaction, would come to be recognized as providing the foundation of "man–machine systems." Minsky, "Steps toward Artificial Intelligence," 28.
11. Fischler, "Machine Perception," 630.
12. Fischler, "Machine Perception," 633.

13. Fischler, "Machine Perception," 631, 639.
14. The automatic captioning through ImageNet-trained object identification neural networks and the generation of synthetic images from descriptions using generative adversarial models are both problems and activities attracting much attention at present. See Christian Szegedy, Wei Liu, Yangqing Jia, Pierre Sermanet, Scott Reed, Dragomir Anguelov, Dumitru Erhan, Vincent Vanhoucke, and Andrew Rabinovich, "Going Deeper with Convolutions," in *Proceedings of the 2015 IEEE Conference on Computer Vision and Pattern Recognition (CVPR)* (2015): 1–9, https://doi.org/10.1109/CVPR.2015.7298594; Scott Reed, Zeynep Akata, Xinchen Yan, Lajanugen Logeswaran, Bernt Schiele, and Honglak Lee, "Generative Adversarial Text to Image Synthesis," in *Proceedings of the 33rd International Conference on Machine Learning* (2016): 1060–9, http://proceedings.mlr.press/v48/reed16.html.
15. Oscar Firschein and Martin A. Fischler, "Describing and Abstracting Pictorial Structures," *Pattern Recognition* 3, no. 4 (1971): 421–43, 423.
16. Firschein and Fischler, "Describing and Abstracting Pictorial Structures," 426.
17. Firschein and Fischler, "Describing and Abstracting Pictorial Structures," 425.
18. Firschein and Fischler, "Describing and Abstracting Pictorial Structures," 438.
19. Peter T. White, "The Camera Keeps Watch on the World," *New York Times,* April 3, 1966.
20. Martin A. Fischler and Robert A. Elschlager, "The Representation and Matching of Pictorial Structures," *IEEE Transactions on Computers* C-22, no. 1 (1973): 67–92, https://doi.org/10.1109/T-C.1973.223602.
21. Fischler and Elschlager, "Representation and Matching," 67.
22. Fischler and Elschlager, "Representation and Matching," 67.
23. Fischler and Elschlager, "Representation and Matching," 75.
24. Fischler and Elschlager, "Representation and Matching," 75.
25. Fischler and Elschlager, "Representation and Matching," 91.
26. Roopika Risam, "What Passes for Human? Undermining the Universal Subject in Digital Humanities Praxis," in *Bodies of Information: Intersectional Feminism and Digital Humanities,* edited by Elizabeth Losh and Jacqueline Wernimont (Minneapolis: University of Minnesota Press, 2018), 39–56, 51.
27. Fischler and Elschlager, "Representation and Matching," 91.
28. Fischler and Elschlager, "Representation and Matching," 68.
29. Fischler and Elschlager, "Representation and Matching," 68.
30. Matthew Turk and Alex Pentland, "Eigenfaces for Recognition," *Journal of Cognitive Neuroscience* 3, no. 1 (1991): 71–86, https://doi.org/10.1162/jocn.1991.3.1.71.
31. Turk and Pentland, "Eigenfaces for Recognition," 71.

32. Turk and Pentland, "Eigenfaces for Recognition," 73.
33. Paul Viola and Michael Jones, "Rapid Object Detection Using a Boosted Cascade of Simple Features," in *Proceedings of the 2001 IEEE Computer Society Conference on Computer Vision and Pattern Recognition, CVPR 2001* (2001): I-511–I-518, https://doi.org/10.1109/CVPR.2001.990517.
34. OpenCV explains that they are using a Lienhart and Maydt–modified version of the Viola-Jones framework. Rainer Lienhart and Jochen Maydt, "An Extended Set of Haar-like Features for Rapid Object Detection," in *Proceedings of the 2002 IEEE International Conference on Image Processing* (2002): 900–903, https://doi.org/10.1109/ICIP.2002.1038171. The documentation for creating cascade classifiers is available in OpenCV, "Haar Feature–Based Cascade Classifier for Object Detection," https://docs.opencv.org/4.5.4/d5/d54/group__objdetect.html.
35. Minsky, "Steps toward Artificial Intelligence," 15.
36. Targeting difference through algorithms that prioritize what Chun terms "homophily," the love of the same, produces a hatred of otherness, linking love and hate into neighborhood-organizing logics found throughout the artificial intelligence methods found at the intersection of these fields. Wendy Hui Kyong Chun, "Queerying Homophily," in *Pattern Discrimination* (Minneapolis: University of Minnesota, 2019), 59–97.
37. Allan Sekula, "The Body and the Archive," *October* 39 (1986): 3–64, https://doi.org/10.2307/778312.
38. Nikki Stevens and Os Keyes, "Seeing Infrastructure: Race, Facial Recognition and the Politics of Data," *Cultural Studies* 35, no. 4–5 (2021): 833–53, https://doi.org/10.1080/09502386.2021.1895252.
39. Wendy Hui Kyong Chun, *Discriminating Data: Correlation, Neighborhoods, and the New Politics of Recognition* (Cambridge, Mass.: MIT Press, 2021), 212.
40. Harry G. Barrow, "Interactive Aids for Cartography and Photo Interpretation," semiannual technical report, contract DAAG29-76-C-0057 (Menlo Park, Calif.: SRI International, 1977), ii.
41. Martin A. Fischler, "Interactive Aids for Cartography and Photo Interpretation," semiannual technical report, contract DAAG29-76-C-0057 (Menlo Park, Calif.: SRI International, 1978), ii.
42. Fischler, "Interactive Aids," 16.
43. Barrow, "Interactive Aids," 1.
44. Martin A. Fischler, "Image Understanding Research and its Application to Cartography and Computer-Based Analysis of Aerial Imagery," sixth and seventh semiannual technical reports covering the period April 1, 1982, to March 31, 1983, contract MDA903-79-C-0588 (Menlo Park, Calif.: SRI International, 1983), 3.

45. Richard O. Duda and Edward H. Shortliffe, "Expert Systems Research," *Science* 220, no. 4594 (1983): 261–68, 261, https://doi.org/10.1126/science.6340198.
46. H. M. Collins, "Expert Systems and the Science of Knowledge," in *The Social Construction of Technological Systems,* ed. Wiebe E. Bijker, Thomas P. Hughes, and Trevor Pinch (Cambridge, Mass.: MIT Press, 2012), 332–34.
47. In his history, McCarthy connects the origins of LISP with the origins of artificial intelligence: "My desire for an algebraic list-processing language for artificial intelligence work on the IBM 704 computer arose in the summer of 1956 during the Dartmouth Summer Research Project on Artificial Intelligence, which was the first organized study of AI." John McCarthy, "History of LISP," *ACM SIGPLAN Notices* 13, no. 8 (1978): 217–23, 217.
48. Beaumont Newhall, *Airborne Camera: The World from the Air and Outer Space* (New York: Hastings House, 1969), 63.
49. Caren Kaplan, *Aerial Aftermaths: Wartime from Above* (Durham, N.C.: Duke University Press, 2018), 8.
50. Kaplan, *Aerial Aftermaths,* 212.
51. Hrishikesh B. Aradhye, Martin A. Fischler, Robert C. Bolles, and Gregory K. Myers, "Method and Apparatus for Person Identification," U.S. Patent 7,792,333 B2, filed October 19, 2005, and issued September 7, 2010.
52. Aradhye et al., "Method and Apparatus for Person Identification."

4. SHAKY BEGINNINGS

1. There was some initial confusion over the correct spelling of Shakey, as SRI International used a nonstandard spelling for their robot and project.
2. Brad Darrach, "Meet Shaky, the First Electronic Person," *Life,* November 20, 1970.
3. Darrach, "Meet Shaky," 58D.
4. Darrach, "Meet Shaky," 58C.
5. Shakey's gendering as male is explicitly invoked in a caption for the last image appearing in the article: "Shaky and one of the computer scientists who helped create him eye each other in an office at Stanford. The scientists off handedly refer to Shaky as 'he,' and one says 'we have enough problems already without creating a female Shaky. But we've discussed it.'" (What exactly makes Shakey male, and what would be involved in creating a female Shakey, are left unsaid.) Darrach, "Meet Shaky," 68.
6. The *New York Times* article does not reference Shakey by name, instead referring to Shakey as "an ungainly automaton" that "moves

slowly and somewhat shakily." Robert Reinhold, "'Baby' Robot Learns to Navigate in a Cluttered Room," *New York Times,* April 10, 1968.

7. Like Shakey, Flakey was developed by SRI International as another autonomous mobile robot, but it implemented a fuzzy logic controller—hence the name. Like Shakey, Flakey was also partly supported by Defense Department organizations, this time the U.S. Air Force and the Office of Naval Research. Alessandro Saffiotti, Enrique H. Ruspini, and Kurt G. Konolige. "A Fuzzy Controller for Flakey, an Autonomous Mobile Robot," Technical Note 529 (Menlo Park, Calif.: SRI International, March 1993).
8. Leonard J. Chaitin, Richard O. Duda, P. A. Johanson, Bertram Raphael, Charles A. Rosen, and Robert A. Yates, "Research and Applications: Artificial Intelligence," NAS12-2221, Technical Report 19730016446 (Washington, D.C.: NASA, April 1, 1970), iii, https://ntrs.nasa.gov/citations/19730016446.
9. Nils J. Nilsson, ed., "Shakey the Robot," Technical Note 323 (Menlo Park, Calif.: SRI International, April 1984).
10. Benjamin Kuipers, Edward A. Feigenbaum, Peter E. Hart, and Nils J. Nilsson, "Shakey: From Conception to History," *AI Magazine* 38, no. 1 (2017): 88–103.
11. Charles A. Rosen, Nils J. Nilsson, and Milton B. Adams, "A Research and Development Program in Applications of Intelligent Automata to Reconnaissance—Phase 1," Proposal ESU 65-1 (Menlo Park, Calif.: Stanford Research Institute, January 1965), 4.
12. Cited in Nils J. Nilsson, "Introduction to the COMTEX Microfiche Edition of the SRI Artificial Intelligence Center Technical Notes," *AI Magazine* 5, no. 1 (1984), 41–52, 42–43, https://doi.org/10.1609/aimag.v5i1.424.
13. For an account of the later developments of the relation between DARPA and SRI International, see Michael Belfiore, *The Department of Mad Scientists: How DARPA Is Remaking Our World, from the Internet to Artificial Limbs* (New York: HarperCollins, 2009), 76–94.
14. Rosen, Nilsson, and Adams, "Research and Development Program," 2.
15. Rosen, Nilsson, and Adams, "Research and Development Program," 11.
16. Kuipers et al., "Shakey," 93.
17. Rosen, Nilsson, and Adams, "Research and Development Program," 3.
18. Nils J. Nilsson, *Quest for Artificial Intelligence,* 69–70.
19. Alfred E. Brain, George Forsen, David Hall, and Charles Rosen, "A Large, Self-Contained Learning Machine," in *Proceedings of the Western Electronic Show and Convention,* August 20–23, 1963, San Francisco, California.
20. Minsky, "Steps toward Artificial Intelligence," 15.
21. Hough would join hundreds of academics, including a number of his Brookhaven National Laboratory colleagues, in signing a statement

demanding the end of all military operations in Vietnam, authored by the "Ad Hoc Universities Committee for the Statement on Vietnam" on behalf of the Committee of the Professions and published in the *New York Times* on June 5, 1966. He does not appear to have made any statements about his algorithm's being modified and used for a military-funded robotics project. "On Vietnam," *New York Times,* June 5, 1966.

22. Nobel Foundation, "The Nobel Prize in Physics 1960," https://www.nobelprize.org/prizes/physics/1960/summary/.
23. Paul V. C. Hough, "Method and Means for Recognizing Complex Patterns," U.S. Patent 3,069,654, filed March 25, 1960, and issued December 18, 1962.
24. Paul V. C. Hough, "A Computer Learns to See," in *Unity of Science: Brookhaven Lecture Series* (Upton, N.Y.: Brookhaven National Laboratory, February 14, 1962), 5.
25. Hough, "Computer Learns to See," 4.
26. Hough, "Computer Learns to See," 4.
27. Hough, "Computer Learns to See," 4.
28. Richard O. Duda and Peter E. Hart, "Use of the Hough Transformation to Detect Lines and Curves in Pictures," *Communications of the ACM* 15, no. 1 (1972): 11–15.
29. Peter E. Hart, "How the Hough Transform Was Invented," *IEEE Signal Processing Magazine,* November 2009, 18–22, https://doi.org/10.1109/MSP.2009.934181.
30. The comparison of values in the accumulating arrays has been described as a voting mechanism by some, but Duda and Hart do not discuss the operation of the Hough transform algorithm in those terms. The slightly earlier k-NN learning algorithm, as glossed by Hart and coauthor Thomas M. Cover, uses this pseudo-democratic logic. Thomas M. Cover and Peter Hart, "Nearest Neighbor Pattern Classification," *IEEE Transactions on Information Theory* 13, no. 1 (1967): 21–27, https://doi.org/10.1109/TIT.1967.1053964. The Wikipedia entry on the Hough transform references the algorithm's "explicit voting procedure" (https://en.wikipedia.org/wiki/Hough_transform).
31. "Before coming to SRI, Dr. Nilsson completed a three-year term of active duty in the U.S. Air Force. He was stationed at the Rome Air Development Center, Griffiss Air Force Base, New York. His duties entailed research in advanced radar techniques, signal analysis, and the application of statistical techniques to radar problems. He has written several papers on various aspects of radar signal processing. While stationed at the Rome Air Development Center, Dr. Nilsson held an appointment as Lecturer in the Electrical Engineering Department of Syracuse University." "Proposal for Research: SRI No.

ESU 68-111," September 4, 1968, https://www.sri.com/wp-content/uploads/2021/12/1291.pdf.

32. Nilsson, "Introduction to the COMTEX Microfiche Edition." In an interview with William Aspray recorded in Palo Alto, California, in March 1989, Nilsson reconfirmed the way in which he believed the Mansfield amendment had altered the nature of work in robotics and artificial intelligence. Nils J. Nilsson, "An Interview with Nils Nilsson" (Minneapolis: Charles Babbage Institute, Center for the History of Information Processing, University of Minnesota, March 1989), https://conservancy.umn.edu/bitstream/handle/11299/107545/oh155nn.pdf.
33. Comptroller General of the United States, "Implementation of 1970 Defense Procurement Authorization Act Requiring Relationship of Research to Specific Military Functions," June 23, 1970, https://www.gao.gov/assets/680/674503.pdf.
34. For an examination of this era focused on weapons research and from the perspective of physics, see Sarah Bridger, *Scientists at War: The Ethics of Cold War Weapons Research* (Cambridge, Mass.: Harvard University Press, 2015).
35. Weldon B. Gibson, *SRI: The Take-off Days* (Los Altos, Calif.: Publishing Services Center, 1986), 6.
36. Stuart W. Leslie, *The Cold War and American Science: The Military-Industrial-Academic Complex at MIT and Stanford* (New York: Columbia University Press, 1994).
37. "Stanford Research Goes to War," *Resistance* 1, no. 2 (1967): 2.
38. "Stanford Research Goes to War," 3.
39. John Walsh, "Stanford Research Institute: Campus Turmoil Spurs Transition," *Science* 164, no. 3882 (1969): 933–36, 934, https://doi.org/10.1126/science.164.3882.933.
40. Walsh, "Stanford Research Institute," 934.
41. Writer and academic Ed McClanahan participated in the protest while teaching at Stanford. McClanahan had initially come to Stanford for the prestigious Stegner Fellowship program (1962–63), then stayed until 1972, teaching under the auspices of the Jones Lectureship. He provides a lightly fictionalized account of the protest in Ed McClanahan, "Another Great Moment in Sports," in *O, the Clear Moment* (Berkeley, Calif.: Counterpoint, 2008), 71–89.
42. Lang Atwood and Frank Miller, "SRI Obtains Injunction; Movement Plans Action," *Stanford Daily,* May 19, 1969.
43. Leslie, *Cold War and American Science,* 247.
44. Richard W. Lyman, *Stanford in Turmoil: Campus Unrest, 1966–1972* (Stanford, Calif.: Stanford University Press, 2009), 121.
45. The timeline of the Computation Center protest and related events provided by the April Third Movement website suggests that Lingg is

a pseudonym (http://www.a3mreunion.org/archive/1970-1971/70-71_laos_computer/1970-1971_laos_computer.html). The guest column is in the *Stanford Daily* archives. Lew Lingg, "SRI War Games at Computation Center," *Stanford Daily,* February 9, 1979, 2, https://archives.stanforddaily.com/1971/02/09?page=2§ion=MODSMD_ARTICLE6#a.

46. A recording of this discussion is available as part of Stanford University's Activism @ Stanford digital collection: "Student Rally in Dinkelspiel Auditorium and White Plaza (From Start of Rally to Departure for Computation Center Sit-in)," February 9, 1971, Stanford University audio collection, circa 1972–2002 (inclusive), https://purl.stanford.edu/bj285nd6494. Lyman's account of this debate over plans and tactics for what was termed a "strike" at the Computation Center names several speakers, including law student Janet Weiss. Lyman, *Stanford in Turmoil,* 182–83.
47. Lyman, *Stanford in Turmoil,* 183.
48. Franklin quoted in Lyman, *Stanford in Turmoil,* 183.
49. Matt Tierney, *Dismantlings: Words against Machines in the American Long Seventies* (Ithaca, N.Y.: Cornell University Press, 2019), 44.
50. Lyman, *Stanford in Turmoil,* 185.
51. Nilsson interview with Aspray, 1989.
52. James Lighthill, "Artificial Intelligence: A General Survey," in *Artificial Intelligence: A Paper Symposium* (London: U.K. Science Research Council, 1973).
53. Lighthill, "Artificial Intelligence."
54. Several studies examine the loss of faith in artificial intelligence and the differing accounts of ARPA's financial support of these technologies in the period after that examined in this book: Daniel Crevier, *AI: Tumultuous History of the Search for Artificial Intelligence* (New York: Basic Books, 1993); Mikel Olazaran, "A Sociological Study of the Official History of the Perceptrons Controversy," *Social Studies of Science* 26, no. 3 (1996): 611–59, https://doi.org/10.1177/030631296026003005; Jon Guice, "Controversy and the State: Lord ARPA and Intelligent Computing," *Social Studies of Science* 28, no. 1 (1998): 103–38, https://doi.org/10.1177/030631298028001004; Pamela McCorduck, *Machines Who Think: A Personal Inquiry into the History and Prospects of Artificial Intelligence* (Natick, Mass.: A. K. Peters, 2004).

CODA

1. The history of Intel's initial development and the opensource release of OpenCV can be found in some of the tool kit's documentation, the website, and on the Wikipedia page for OpenCV (https://en.wikipedia.org/wiki/OpenCV).

2. The documentation for these major wrapped algorithms as methods, which include HoughLines, HoughCircles, SimpleBlobDetector, matchTemplate, CascadeClassifier, and ml.ANN_MLP_create, can be found on the OpenCV documentation site (https://docs.opencv.org).
3. "About," OpenCV, http://www.opencv.org/about.
4. "LS3 Pack Mule," DARPA, https://www.darpa.mil/about-us/timeline/legged-squad-support-system.
5. The "Choreography" module of Spot's API contains ready-made functions to perform *pas de bourrée,* a chicken head bob, the Running Man, and a function called ButtCircle, among others (https://dev.boston dynamics.com/choreography_protos/bosdyn/api/readme).
6. Atanasoski and Vora, *Surrogate Humanity,* 145.
7. "Spot," Boston Dynamics, https://www.bostondynamics.com/products/spot.
8. Boston Dynamics, "Spot Launch," YouTube, September 24, 2019, video, 2:01, https://www.youtube.com/watch?v=wlkCQXHEgjA.
9. Harun Farocki, "Phantom Images," *Public* 29 (2004): 12–22, 17.
10. Farocki, "Phantom Images," 17.
11. Trevor Paglen, "Operational Images," *e-flux* 59 (2014): 1–3, 1, https://www.e-flux.com/journal/59/61130/operational-images/.
12. Paglen, "Operational Images," 3.
13. Virilio, *Vision Machine,* 60.
14. Fabian Offert and Peter Bell, "Perceptual Bias and Technical Metapictures: Critical Machine Vision as a Humanities Challenge," *AI and Society* 36 (2021): 113–33, 1135–36, https://doi.org/10.1007/s00146-020-01058-z.
15. Jenna Burrell, "How the Machine 'Thinks': Understanding Opacity in Machine Learning Algorithms," *Big Data and Society* 3, no. 1 (2016): 1–12, 6, https://doi.org/10.1177/2053951715622512.
16. Burrell, "How the Machine 'Thinks,'" 3.
17. Hito Steyerl, "A Sea of Data: Pattern Recognition and Corporate Animism (Forked Version)," in *Pattern Discrimination* (Minneapolis: University of Minnesota Press, 2019), 10.
18. Luciana Parisi, "Negative Optics in Vision Machines," *AI and Society* 36 (2021): 1281–93, 1283, https://doi.org/10.1007/s00146-020-01096-7.
19. Parisi, "Negative Optics," 1286.
20. Farocki, "Phantom Images," 18.

Index

ADALINE, 142
aerial imagery, 31, 40, 47–48, 58–59, 76, 82–84, 89, 107–11, 117, 125–26, 153, 169–70
aerial photography, 129–30
affordances, 12, 14, 21, 23, 49, 52
Agent Orange, 118
Air War Study Group, 95–96
algorithm, 2–8; defined, 185n40
Amazon, 9, 15, 181n26; Alexa, 9
android, 12
Apple, 9; Siri, 9
artificial intelligence, 11–12, 16–17, 20, 25, 38–39, 40–43, 61–63, 90–95, 103–4, 123–25, 128, 131, 133, 135–37, 142, 151, 161–62, 180n18, 181n27, 184n23, 186n1. *See also* machine learning
automata, 137–40
automatic programming, 77
automaton, 13, 24–25, 82–83, 90–91, 137–39, 141, 187n13, 194n6
averaging, 5, 36, 87, 100

Bayes's rule, 76, 104
Bazin, André, 34
BBC, 79–80
bias, 3, 10, 12, 16, 20, 37, 48, 100
bird's-eye perspective, 7, 59, 129, 133
blob detection, 85, 110, 166, 176
blur, 36, 87
body, 7–8, 11, 120, 123
Bogost, Ian, 52–53
Boston Dynamics, 167–69
bounding box, 46, 54, 119, 120, 121, 169, 170
Brookhaven National Laboratory, 14, 147–49
Browne, Simone, 16, 181n26
bubble chamber, 147–49
Buolamwini, Joy, 100
ButtCircle, 199n5

Calspan Corporation, 64, 186n6
camera obscura, 4
captioning, 102–3, 107, 191n7, 192n14
cell assemblies, 187n23
Chun, Wendy, 124, 193n36
classification, 11, 25, 34, 35, 51, 55, 57, 61, 74, 76, 78, 80, 86, 92, 94, 103, 109, 123–24, 144, 172; algorithms, 54, 56, 104, 121; gender, 79–80; image, 25, 77, 103, 144, 172; linear, 23; problem defined, 53, 55–56
Cold War, 10, 21, 24, 27, 51, 61, 64, 74–75, 83, 97
command and control systems, 12
computer graphics, 23, 30–32, 171, 183n11
conditional probability computer, 76
Connection Machine, 189n60
convolutional neural networks (CNNs), 25, 63, 172

Cornell Aeronautical Laboratory (CAL), 14, 23, 53, 63–64
Cornell University, 23, 64, 66, 76, 77, 83, 95, 155, 156, 162; cognitive systems research program, 64, 78
Crary, Jonathan, 4–5
creative writing, 107
critical algorithm studies, 8–11
critical code studies, 48–49
Curtiss-Wright Aircraft Corporation, 63, 186n4
cybernetics, 13, 20–21, 41, 51, 150, 182n2

Dartmouth College, 45, 62, 186n1
Dartmouth Summer Research Project on Artificial Intelligence, 62, 128, 194n47
deep neural networks, 6, 33, 63
Defense Advanced Research Projects Agency (DARPA), 24, 25, 28, 156, 162, 167, 183n5, 189n60, 195n13
Digital Equipment Corporation (DEC), 14, 137
digital objects, 30, 53, 56, 172; modeling, 2, 8, 31, 41; recurrent causality, 30, 103; unit operation, 52–54, 56
discorrelated images, 178n8
discrimination, 34, 53; linear, 62. *See also* classification
drones, 5, 59, 109, 127, 130, 153, 177n5
Duda, Richard, 16, 17, 55, 56, 128, 141
dynamic programing, 181n24

Edwards, Paul, 74, 184n23
Eigenfaces, 118–20, 166
Eisenhower, Dwight, 28
Elschlager, Robert, 109–13, 115, 117, 118, 119
ENIAC, 66
EPAC, 66
excitation, 44, 66, 68, 71, 77
exclusive or (XOR) function, 94
expert systems, 124, 125, 127–28

face detection, 110, 115, 118–21, 123
feature extraction, 35, 52, 57, 89, 145, 155
Firschein, Oscar, 40–42, 106–8
Fischler, Martin, 40–43, 103–18, 125–27, 131, 181n24
Fisher, Ronald, 74, 186n2
Flakey project, 137, 195n7
Foucault, Michel, 22
Franklin, Bruce, 160–61

Gaboury, Jacob, 31–32, 184n25
Galloway, Alexander, 182n31
GAMUT-H, 159
genealogical analysis, 4, 11, 20, 21–22, 25, 49, 62, 110, 176
general artificial intelligence, 24, 135
general-purpose computers, 12–13, 23, 66, 75, 83, 90, 138
generative adversarial models, 25, 33, 183n14, 192n14
Golumbia, David, 50
Google, 9, 167, 173, 175, 181n22; Arts & Culture app, 175

Haar-like features, 119–20
Halpern, Orit, 21
Haraway, Donna, 7–8
Hart, Peter, 125, 137, 140–41
Hawkeye project, 126–28, 131, 133, 135
Hebb, Donald, 70–71, 187n23
Holmes, William, 53–55, 82–89, 109–10, 125
Hough transform, 21, 146–54, 166, 176

Hui, Yuk, 30, 32, 53, 56, 103
human–computer interaction, 9, 119

IBM, 13, 14, 66, 79, 83, 90, 156, 187n23
image concept, 43–48
ImageNet, 191n4, 192n14
image processing, 6, 29, 40, 167
image understanding, 122–32
instrumental vision, 7
isomorphism, 6, 32, 44

Kaplan, Caren, 130
k-nearest neighbors (KNN), 54, 104, 196n30
Kolmogorov-Smirnov filter, 86–87
KZSU, 159

lane departure warning systems (LDWS), 135, 153, 154
lenticular modularity, 51
Life, 135–36
Lighthill, James, 162
line analysis, 140
linear classification, 23, 62, 74, 75, 79, 94, 104, 116
linear embedding model (LEA), 116
line segment analysis and description program (LSADP), 104–6
LISt Processor (LISP), 128, 194n47
Lockheed Corporation, 103, 106, 109
Lyman, Richard, 158, 198n46

machine learning, 2, 6, 11–13, 16, 20, 24, 27, 31, 39, 42, 53–54, 58, 62, 66–72, 75–82, 91, 93, 95, 97–102, 116, 121, 123, 135, 138, 142–46, 150, 152, 162, 166, 172, 173. *See also* artificial intelligence
machine vision, 3, 4, 41, 104, 131, 174

Manovich, Lev, 31, 59
Mansfield amendment, 95, 155, 163, 197n32
Marino, Mark, 48
Mark I, 66, 69, 75–78, 79, 83, 90, 98, 142
Massumi, Brian, 12
Maturana, Humberto, 71
McCarthy, John, 62, 91, 128, 186n1
McCulloch, Warren, 70–71
McPherson, Tara, 51
metapicture, 172
Microsoft: corporation, 177n4; office, 102, 191n8
Miller, Arthur, 186n4
Miller, Rodman, 190n71
MINOS, 76, 142
Minsky, Marvin, 62, 91, 92–96, 98, 120, 135, 136
MIT, 38, 91, 135, 155, 156, 167; Media Lab, 118; Project MAC, 135, 184n23
Mitchell, W. J. T, 172
MITRE Corporation, 156
mobile frame, 5, 50, 142, 153, 169
multivariate analysis, 64
Murray, Albert E., 76, 81, 82

neural networks, 6, 25, 33, 39, 44, 54, 63, 70, 94, 95, 171–73
neurodynamics, 93
neuroscience, 21, 40
New Yorker, 90
Newhall, Beaumont, 129
Nilsson, Nils, 139, 140, 154, 155, 161, 162, 181n27
normative bodies, 21, 112–18, 132

object detection, 12, 85, 97, 111, 114, 119, 153
Office of Naval Research (ONR), 76, 79, 82, 83, 107
ontology, 5, 6, 10, 13, 21, 24, 30, 33–38, 53, 54, 56, 71, 87, 98,

100, 102, 122, 134, 146, 153, 171, 178n8
ontopower, 12
opacity, 43, 172–73, 177n2
OpenCV, 25, 120, 152, 153, 165–67, 169, 181n22, 185n38, 193n34
operative image, 25, 170–73
optical character recognition, 59, 162

panopticon, 7
PAPA, 77
Papistor, 77
pattern recognition, 16, 21, 23, 28, 35, 38–40, 57, 76–78, 82, 85, 92–94, 110–12, 124, 135, 144, 148–49, 169, 176
perception, 2–4, 7–9, 12–17, 19, 21, 27, 29, 33, 37, 43, 50–51, 56, 58, 60, 63, 67–68, 70–72, 75, 78, 97–100, 104, 106, 122–23, 133, 135–36, 142, 161, 171–72, 174–75
Perceptron, 21, 23, 24, 44, 61–81, 83, 89–96, 97, 98, 99, 109, 135, 142, 144, 166, 176
phantom photographer, 36–37
phenomenology, 9, 53, 178n13
photography: aerial, 5, 24, 34, 58, 59, 109, 111, 129, 130, 131, 153; digital, 33, 179; film, 4, 36, 58, 110, 123; surveillance, 48
photointerpretation, 24, 63, 79, 81–84, 123–32, 133
pictorial structures, 21, 103, 108, 109–12, 118, 119, 134, 144, 176
picture primitives, 108, 113, 114, 116
picture processing, 1, 31, 32, 43, 166
pixels, 6, 35, 43–47, 54, 55, 71, 72, 84–88, 98, 101, 103, 109, 112, 119, 122, 144–45, 152, 169
platform seeing, 50
policing, 1, 2, 11, 16, 48, 50, 158, 179n16
progress myth, 11, 15, 135, 180n18
Project PARA, 67, 78, 79
property derivation, 53–55
psychology: behavioral, 64–67, 72, 95, 104; gestalt, 71; social, 64

quantization, 33, 43, 46, 53, 79, 86, 98, 101
quick response (QR) codes, 10

realism: literary, 107; photographic, 34, 35
reconnaissance imaging, 24, 37, 39, 40, 51, 59, 63, 83, 109, 117, 125, 129, 137, 157, 170
remediation, 18, 19, 22, 49
Reserve Officers' Training Corps (ROTC), 156
Resistance, 157, 160
Roberts, Lawrence, 38, 184n25
Rome Air Development Center, 137, 153, 196n31
Rosen, Charles, 78, 80, 81, 135, 138, 139, 140, 141, 142, 161
Rosenblatt, Frank, 23–24, 64–73, 75–79, 81, 83, 86, 89–98, 109, 135, 142, 144, 162
Rosenfeld, Azriel, 27, 31, 32, 39, 40, 151
Rouvroy, Antoinette, 177n2

Scientific Data Systems, 14, 137, 138
segmentation, 18, 84–89, 131, 146
Sekula, Allan, 123
selfie, of author, 121
self-organizing systems, 13, 41, 83, 150
sensed scenes, 32, 37, 100, 111, 145
Shakey Project, 134–43, 146–47, 151, 153, 155, 158, 167–69, 171, 194n1

Shannon, Claude, 62, 140
Shearwater, 95
sightless seeing, 6, 25, 37, 50, 122, 169, 170, 171, 173, 174
Simondon, Gilbert, 30
simulation, 30–31, 40, 41, 42, 70, 79, 82, 93, 95, 122, 159, 169
smartphones, 9, 10, 15, 50, 60, 154, 175, 184n22
social construction of technological systems, 9, 191n6
Sontag, Susan, 34
speech recognition, 77–78, 85
Spot (robot), 167–69
SRI International, 14, 24, 40, 41, 125, 126, 131, 134–36, 138, 141, 150, 154, 156, 158, 161–62, 167
Stanford University, 138, 142, 155–58, 162; computation center, 159–61
surveillance, 1, 2, 7, 9, 12, 24, 31, 39, 40, 46, 48, 50, 59, 60, 63, 78, 87, 89, 100, 111, 116–17, 121, 123, 125, 129, 130–34, 153, 155, 157, 166–69, 172, 175
surveillance capitalism, 2, 50, 177n4
surveillance culture, 61, 83
symbolic representation, 6, 52, 55, 68, 97, 101–3, 106, 110, 128, 145

technical objects, 12, 30. *See also* digital objects
template pattern matching, 21, 87–89, 92, 109–12, 116, 176
tensor processing unit, 181n22
Thinking Machines Corporation, 189n60
Tobermory, 77–78, 183n6
tool criticism, 178n12
training, 2, 20, 35, 36, 53, 54, 57, 62, 69, 76, 79, 80–84, 102, 107, 123, 128
transparency ideal, 178n13
Turing, Alan, 70
TV camera, 78, 104
Twain, Mark, 107

U.S. Navy, 23, 91, 93
Uttley, Albert, 76

vector representations, 16. *See also* digital objects
Vietnam War, 12, 24, 64, 109, 117, 128, 131, 134, 155, 157, 158, 162, 195n21
Viola-Jones, 119, 166, 193n34
Virilio, Paul, 37, 170–71, 174
vision models: cat, 63, 78, 186n4; frog, 42, 63, 71, 72; human, 71

war games, 159
Winner, Langdon, 180n18, 191n6
Winston, Patrick Henry, 40, 42
Wolf, Helen, 141
workflow, 53–57, 107, 122, 144, 152, 170, 172

Žižek, Slavoj, 181n25
Zuboff, Shoshana, 2

James E. Dobson is assistant professor of English and creative writing and director of the Institute for Writing and Rhetoric at Dartmouth College. He is author of *Modernity and Autobiography in Nineteenth-Century America: Literary Representations of Communication and Transportation Technologies* and *Critical Digital Humanities: The Search for a Methodology* and coauthor of *Moonbit.*